From The Vault of Schrodinger's Cat

A Collection of Posts From the Schrodinger's Cat Blog Featuring Updated Footnotes

Natasha Parkinson

First paperback edition July 2023

Book design by Natasha Parkinson

ISBN 978-1-7390245-0-5

Published by Natasha Danielle: The STEM Gal Media
thestemgalmedia.com

This book is dedicated to

my Dad, who was always proud of my big "sciencey" brain and never missed an opportunity to call me his "little Sheldon Cooper."

my Mom, who had to listen to all my random science facts when I was growing up and helped me put together my first science fair project when I was in Grade 2.

& Nick for showing me the "wayback" to Schrodinger's Cat.

Table of Contents

Health Science

Math + Physics

Science In Pop Culture

Space

Technology

Works Cited

<u>Foreword</u>

Schrodinger's Cat was a project that arose after I caught myself countless times fact-checking my friends' posts on social media. I realized that social media was rampant with misleading information and pseudoscience, which is difficult for the average person to sort out. It was through my training from some of the most esteemed professors that I have had the pleasure to be their student, that I learned to pull facts from fabrication and ask questions like, "What is their motivation? Who is backing this? Is it information or an infomercial?" This is a skill set that I wish everyone had the opportunity to receive in their educational journey. Unfortunately, that is not the case. So it became my mission through Schrodinger's Cat to provide informative content that my readers could trust.

Each section in this book is a blog post that at one time was a part of the Schrodinger's Cat blog, just slightly reformatted. As such, you may notice the in-text citation is extensive, almost to excess. This is because, in the original format, I had hyperlinks for my readers to be able to read the articles that I got my information from (in the spirit of 100% transparency). For obvious reasons, it's difficult to have hyperlinks in a book, especially in a physical copy. As a result, the passages that had hyperlinks now have an in-text citation, with full citations listed in the Appendix.

Biochemistry + Chemistry

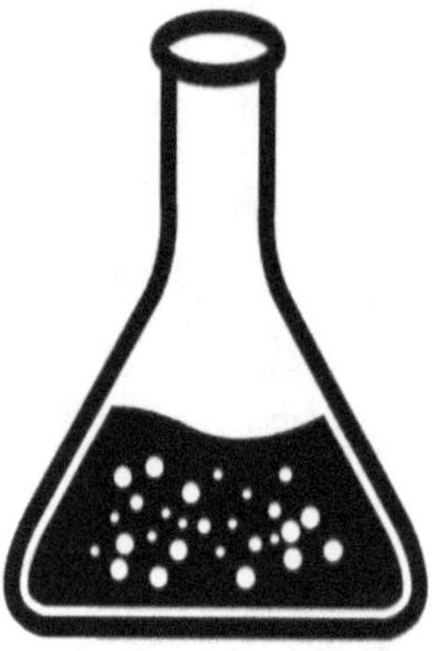

The Chemistry of Cannabis

(Originally Published June 6, 2018)

Last week, Canada passed new legislation making it the second country ever to nationally legalize marijuana. As there are many people that are overjoyed by the news, there are also some that are reluctant to this new change. There is plenty of time between now and October 2018 (when legalization is scheduled to occur in Canada), so let's take this opportunity to get some education out about marijuana and debunk some of the myths that are floating around out there. So, let's get started on the Countdown to Cannabis, Part 1, and talk about some of the differences in the chemistry of cannabis.

Marijuana use by humans dates back as far as the Neolithic period (8,000 – 5,000 BC) ("Inside: Marijuana Facts," n.d.). There are two main species of the cannabis plant that humans utilized, *Cannabis indica* and *Cannabis sativa* (NIDA, n.d.). There are two main active chemical components of marijuana, THC and CBD (although, there are more than 100 different cannabinoids present in marijuana plants), that vary in concentration depending on the species of plant. Indicas have a higher level of THC, whereas, Sativas have a higher level of CBD (Hilig et al., 2004). Because of these differences, these plants have varying effects on those that use them. Indicas are best suited for those that are seeking to relax or calm down and are used by those dealing with conditions such as MS, glaucoma, chronic pain, Crohn's disease, anxiety, and troubles with sleeplessness (Rafael-Hawkins, 2017). Whereas, Sativas are more energizing and used by those dealing with mental ailments, such as anxiety, ADD/ADHD, depression, and mood disorders (Rafael-Hawkins, 2017). In addition to these two main species, there are also hybrids of both Indicas and Sativas, in order to achieve a desired THC and CBD blend concentration. But what are THC and CBD?

THC

Tetrahydrocannabinol, or THC, is the chemical in marijuana that gives the "high" feeling (Vandergriendt, 2019). It is because of this psychoactive effect that many of the stigmas against marijuana users come from, as well as, why Canada's legalization of marijuana has been a long process. As the legalization process moves forward, many law enforcement branches are trying to figure out a way to test for those under the influence, as the traditional means of testing just indicates if an individual has used marijuana in the past 21 days or so (as the time to clear the body varies from individual to individual). Despite the impairment that THC can cause, it does have many benefits, including pain management, when used correctly, without the harsh side effects of many pharmaceutical methods.

THC

CBD

Cannabidiol, or CBD, has nearly no psychoactive, or "high," effect (Vandergriendt, 2019). Many that utilize the purified form of CBD oil, you wouldn't even be able to point them out on the street, as they do not exhibit any of the stereotypical "high" characteristics, and most even experience an improved ability to function in their day-to-day lives. Unlike many countries, including most of the United States, CBD oil has been legal in Canada for quite some time and has been used by many to treat various ailments without a prescription.

CBD

The Downside

Despite the benefits that marijuana, THC, and CBD present, there are some drawbacks. First of all the psychoactive effects, or "high" symptoms, that THC can present. These can include altered senses, altered sense of time, mood changes, impaired body movement, difficulty with critical thinking and problem solving, reduced memory, hallucinations, delusions, paranoia, and psychosis (NIDA, n.d.). Also, for those who start to use at a younger age, in their teenage years, there have been cases of impaired brain development (NIDA, n.d.). However, these risks are not unlike those associated with drinking alcohol. Except you rarely hear of a raging pothead, do you?

With the legalization of marijuana in Canada, it presents an opportunity to educate ourselves on alternative therapies. Nothing is without its risks; even vitamin supplements and over-the-counter painkillers have their short-term and long-term risks. And as we approach the big day in Canada, in October 2018, we will look at more information and some of the myths associated with marijuana. So whether you are a Canadian who chooses to partake in the enjoyment of marijuana or not, we hope to have you fully informed when you make your decision. No smoke and mirrors, no bribes from pharmaceutical companies to sway you, just the facts. And that starts with what we discussed today, and how marijuana may have a legal place in society. With that said, thank you, everyone, for tuning in, and be sure to Keep Looking for the Science in Things.

<u>Footnote:</u>

According to Global News, as of February 2022, the cannabis industry in Canada had contributed $43.5 billion to Canada's overall GDP and an additional $15.1 billion to federal tax revenue (The Canada Press, 2022). The booming industry has also created around 151,000 jobs (The Canada Press, 2022). In just two years after legalization (2020), cannabis-related drug offences made up only 19% of all drug-related offences in Canada, whereas prior to legalization cannabis-related drug offences made up 68% of all drug offences (Research and Statistics Division, 2022). That being said, drug-impaired driving rates in Canada have increased by 105% from 2017 to 2020 (Research and Statistics Division, 2022). However, according to the Department of Justice, this may be due in part to new means of detecting drug-impaired driving as a result of new legislation that was brought into effect with the legalization of marijuana.

The Science Behind Fireworks

(Originally Published June 28, 2018)

Over the next week, both Canada and the United States will be celebrating Canada Day and the Fourth of July, respectively. And you know what that means? Besides the fact that everyone has a reason not to have to go to work, there will be FIREWORKS. I don't know about all of you, but I have always been awestruck by fireworks. The pretty colours, the twinkly lights, and the noises. I love fireworks so much that, on more than one occasion, I asked my dad for a fireworks display for my birthday. Fortunately for me, I grew up out in the country, and my birthday is in December. But how do fireworks work? What makes fireworks flash in different colours and even make them come alive? You're about to find out in today's post.

History of Fireworks

Historians have reason to believe that the origin of fireworks came from China some time during the second century B.C., when people threw bamboo stalks onto the fire as a primitive firecracker, which would then produce a loud bang in order to ward off evil spirits (Cohen, n.d.). It wasn't until between 600 and 900 A.D. that Chinese alchemists, searching for the elixir of life, unknowingly produced gunpowder (Cohen, n.d.). With this discovery, Chinese people began to stuff gunpowder into bamboo stalks to form firecrackers (Cohen, n.d.). From there, bamboo was replaced with paper, and the use of firecrackers spread from scaring away evil spirits to celebrating festivals to even being used as crude bombs during military skirmishes (Cohen, n.d.). By the 1400s, gunpowder had made its way into Europe and the Middle East, and along with it, fireworks. Fireworks then found use in celebrating military victories, public celebrations, and religious events (Cohen, n.d.). Colourful and elaborate displays first made an appearance during the Renaissance, with the Italian firework masters discovering that certain metals could be added to give the fireworks different colours (Cohen, n.d.). As time has progressed, fireworks have become the favourite of many European monarchies, hitched a ride to the New World, celebrated the independence of a nation, and even seen a restriction put into place (due to a rampant firework detonation problem). And yet, to this day, fireworks continue to delight millions - not too bad for an accidental discovery.

Making a Firework

Most fireworks manufacturers keep their recipes in absolute secrecy. However, every firework has the same basic construction: the shell, lift charge, black powder, lift charge, a binding resin, a time-delayed fuse, bursting charge gunpowder, and pyrotechnic stars (Dematto, 2010).

> **Shell**, commonly made of cardboard and heavy paper, holds all the ingredients of the firework together as it is launched into the sky (Dematto, 2010).

> **Lift Charges** are the initial fuse that is lit by the firework technician (Dematto, 2010). It ignites the black powder as well as its time-delayed fuse.

Black Powder is made up of approximately 70% potassium nitrate, otherwise known as saltpetre; 15% charcoal; and 10% sulfur (Dematto, 2010). This combination is responsible for launching the firework off the ground and into the sky.

Binding Resin, typically dextrin, is used to hold all the ingredients together and prevent them from crumbling (Dematto, 2010). This is part of the reason why old fireworks aren't the best to launch. They either end up being rather wimpy or downright dangerous, and they can detonate before they reach a safe height.

Time-Delayed Fuse, similar to those found on dynamite and other explosives, dictates the length of time between initial ignition and when the firework explodes (Dematto, 2010). The longer the fuse, the longer the time until the explosion, and in this case, the longer the firework soars into the sky and the higher up the explosion occurs.

Bursting Charge Gunpowder is the gunpowder at the core of the firework that gets the colourful show going (Dematto, 2010). Once it ignites, it starts the change reaction, which blows apart the shell and ignites the pyrotechnic stars.

Pyrotechnic Stars, or **stars**, are pellets that are embedded into the shell of the firework in various configurations and patterns (Dematto, 2010). These different configurations and patterns allow the fireworks to appear as starbursts, flowers, smiley faces, etc., all based on how they are placed in the shell. In addition to their configuration, the stars vary in shape and size (which dictates how each stream of coloured light will trail out) and contain different metals (which produce different colours). They are made up of five general components:

Fuel – allows the star to burn.

Oxidizer – a chemical compound that produces oxygen, which helps burn the fuel.

Colour-producing chemicals – determine the colour of the star as it burns and finally explodes.

Binder – holds the pellet together and prevents it from crumbling or breaking apart prior to the explosion.

Chlorine Donor – provides chlorine, which helps to make the colour brighter, stronger.

All the Pretty Colours

As previously discussed and as we all know, fireworks appear in a large assortment of colours, and that is because of the colour-producing chemicals found in the pyrotechnic stars. At the base of these chemicals are metals. If you aren't aware, different elements, and by extension, different metals, display different colours as they burn. Please do not try to burn different types of metals at home just to see these colours. Yes, I am a pyromaniac at heart and have tried this myself. Some metals produce toxic gases, which are hazardous. Also, fire is dangerous. So even though I know some of you will try this despite my warning, I AM NOT ENDORSING THIS, and therefore will not take any responsibility for anything that results from you trying to burn stuff to see pretty colours. And for those of you who ignore my warning, Safety First.

Ok, back to the discussion on different metals producing different colours. I find this information is best shown in a chart:

Colour	Metal	Chemical Compound
Red (Intense)	Strontium	$SrCO_3$ (Strontium carbonate)
Red (Medium)	Lithium	Li_2CO_3 (Lithium carbonate) & LiCl (Lithium chloride)
Orange	Calcium	$CaCl_2$ (Calcium chloride)
Yellow	Sodium	$NaNO_3$ (Sodium nitrate)
Green	Barium	$BaCl_2$ (Barium chloride)
Blue	Copper halides	$CuCl_2$ (Copper chloride)
Indigo	Caesium	$CsNO_3$ (Caesium nitrate)
Violet	Potassium	KNO_3 (Potassium nitrate)
Violet-Red	Rubidium	$RbNO_3$ (Rubidium nitrate)
Gold	Iron, Charcoal, or Lampblack	
White/Silver	Titanium powder or Aluminum powder	
White	Beryllium powder	
White (Very Bright)	Magnesium powder	

So, as you can see, there are many different chemical combinations out there to give fireworks their different colours.

Firework Effects

In addition to being able to make fireworks appear in different colours, there are different things that make them glitter, change colour part way through, and seem to never end. Roman candles are famous for being glittery. This is achieved by the type of stars embedded in the shell as well as where they are positioned (Dematto, 2010). Also, little strobes are embedded inside the tube to create a glittering golden feather effect (Dematto, 2010). Fireworks that change colour part way through by using two different types of stars, with one colour at the core of the shell and the second colour stars in the outer layers of the shell (Dematto, 2010). Finally, fireworks that never seem to end, or those that are commonly seen in the grand finale. This is because the shell is made to break into smaller sections, each with a time-delay fuse of varying lengths, so they detonate at different intervals of time. The first explosion breaks these pieces apart and spreads them out in the air, then the following explosions are each section, thanks to their time-delayed fuses.

There is a lot of thought, planning, and science that goes into each individual firework that explodes in a matter of seconds. But that is all part of the beauty and splendour of fireworks. And without the work of early Chinese and Italian scientists, Canada and the United States would not be getting their elaborate display in the upcoming week. So with that, I wish everyone a Happy Canada Day and a Happy Fourth of July. Be safe, have fun, and most importantly, enjoy those fireworks. Keep Looking for the Science in Things

The Colours of the Wind

(Originally Published September 5, 2018)

The children are back in school, and Pumpkin Spice has returned to local coffee shops. It's already September, which means Fall (or Autumn, depending on your preference) is on its way here. And I will be the first to admit that I am a stereotypical female. Fall is my favourite time of year. It's not because of the PSL (Pumpkin Spice Lattes); in fact, I've never actually tried one (oh, for shame), and it's not because I get to pull my boots back out of the closet, although boots, sweaters, and scarves become my wardrobe staples this time of year. It's because of all the colourful leaves. The changing of the leaves, how they dance and swirl as they sail towards the ground, and the sound of them crunching under my boots. It's also no surprise the reason that I love Fall is due to one of nature's most beautiful chemistry shows.

Well, it has to do with the various chemical processes that occur with the changing of the seasons. During Spring and Summer, leaves are luscious green thanks to the abundance of chlorophyll. Chlorophyll is a chemical in the leaves that helps the plant photosynthesize, making food for the plant. The process of photosynthesis is when the plant uptakes carbon dioxide and water and uses the energy absorbed from the Sun, using its chlorophyll, and transforms it into carbohydrates (sugars and starches) and oxygen (which is "exhaled" by the plant) (Bassham and Lambers, n.d.).

Chlorophyll A

In addition to chlorophyll, the plant's leaves also contain chemicals
called carotenes and xanthophyll pigments, which create yellow, orange,
and red colouring (Palm Jr., n.d.). Throughout Spring and Summer, we
do not see these colours because the green from the chlorophyll covers
them up. A sunflower plant is an excellent example of the masking
ability of chlorophyll. Sunflower petals (which are modified leaves) are
yellow and orange due to their lack of chlorophyll, whereas their leaves
and stem are often a vibrant green due to the presence of chlorophyll.

β-Carotene

Xanthophyll

As Fall approaches, the days get shorter, the temperature gets cooler, and
many plants (seasonal and deciduous) start to go dormant. They stop
producing food, and they slow down their functions in preparation for
Winter (e.g. sap, the nutrient highway in plants, stops flowing). Annual
plants die off, while perennials, those that return every year, retreat to
their nutritional storehouse, either tubers or bulbs or the trunk and

branches of the tree, allowing its extensions to die off. In deciduous trees and shrubs, the chlorophyll breaks down in the leaves, which causes the green to fade out and the yellow and orange for the carotenes and xanthophyll pigments to pop (Palm Jr., n.d.). Some trees and plants also carry out other chemical changes in their leaves that result in the production of anthocyanin pigments, which give the leaves a red, purple, or orange colour (Palm Jr., n.d.). Those that are brown still have some chlorophyll residue mixed with the other pigments to provide their colour.

Anthocyanin

Once all the resources are pulled from the leaves back into the tree or plant, a layer of cells forms at the base of the leaf stem, which cuts the leaf's support tissue from the branch and seals off the tree, like a scab or scar (Palm Jr., n.d.). Once the leaf falls, there is a leaf scar left behind on the branch where this cut has occurred.

Why Does It Seem Like the Colours Vary Every Year?

Like many chemicals, carotenes, anthocyanins, and xanthophylls are sensitive to external conditions. In the case of these colourful leaf pigments, they are affected by temperature, sunlight availability, and the amount of water available. Noticed a tremendous red in the trees? Anthocyanin is best produced in temperatures just above freezing, and yet, the vibrancy can be weakened with early frost (Palm Jr., n.d.). Everything so bright and colourful? Heading out with an umbrella on rainy and overcast days to see the leaves or driving to areas that receive a higher annual rainfall may allow you to see a higher intensity of colours as the water increases the potency of the pigments.

Whether you live in the city or out in the country, I challenge you to set out and marvel at all the colours that come with Fall. Maybe you'll find a pretty leaf on the ground to take home to display, or perhaps you will take time to try and catch the leaves as they fall (it's not as easy as it looks), or when no one is looking, rake the leaves into a pile and jump in them like you did when you were a child. The colours of Fall are a beautiful display of nature and her chemistry skills, which is just a part of the everyday science that surrounds us. In the meantime, Keep Looking for the Science in Things (especially everyday Science).

Biology

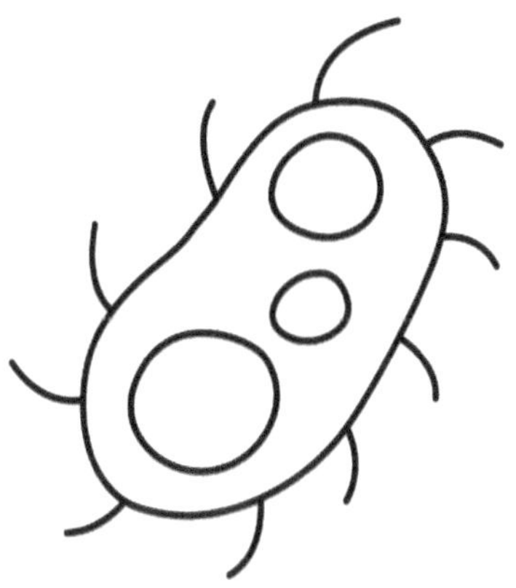

Bet You Didn't Know That Was A GMO

(Originally Published April 18, 2018)

I remember the first time I heard about GMOs when I was a kid. I had this mental picture of an apple crossed with a tuna fish for some reason, and I remember telling my mom all about this apple that had DNA from a fish in it. Of course, she was driving the car at the time, and she asked me what show I got it from because, at that age, I had a very active imagination, and sometimes the line between fact and fiction got blurred a little, sometimes a lot.

But that's the common image that many of us, as adults, get when we hear "GMOs". Some Frankenstein of a fruit or vegetable crawling out of a lab somewhere and wreaking havoc. Is there some validity to those fears? Possibly, and also, possibly not? Let's talk about it.

GMO stands for Genetically Modified Organism, which could be some organism that is created in a lab from gene splicing or simply crossbred or selectively bred by humans in order to amplify certain desirable traits (Zhang et al., 2016).

So, this is a GMO

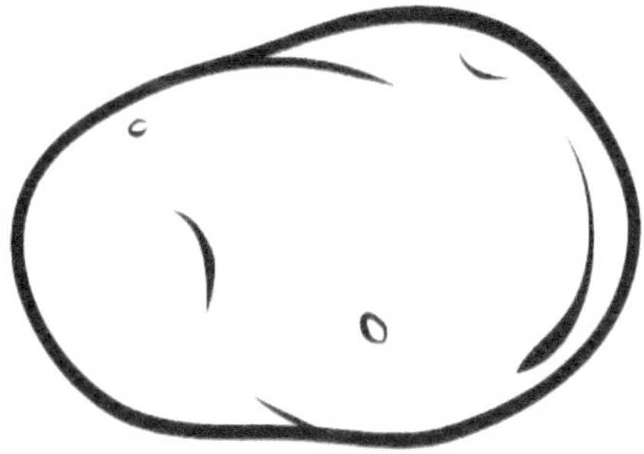

This is a GMO

And these are GMOs

One of the first known GMOs was selectively bred around 10,000 years ago (Olson, 2015). We know it as corn. Prior to human intervention, corn was actually a grass known as teosinte. Through multiple generations of selective breeding and crossbreeding, we came to have corn as we recognize it today. And corn isn't the only GMO that we have due to this method, known as artificial selection. Wheat, rice, almonds, and bananas are also some of the foods that we have today because of artificial selection.

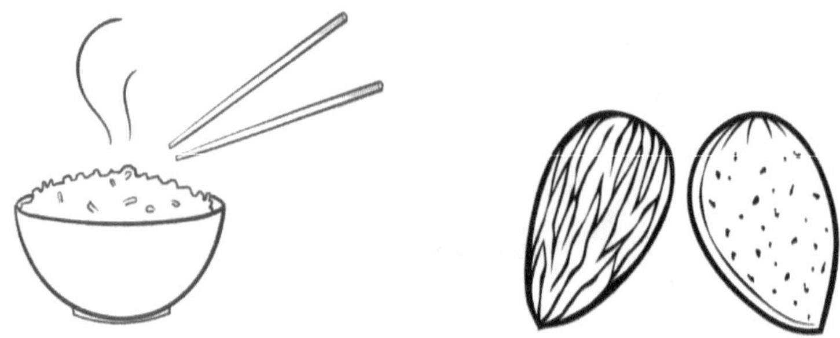

Now, jump ahead to the past 20 years, society has made leaps and bounds in the realm of GMOs by utilizing lab technology in order to speed up the artificial selection process. In simple terms, scientists take an isolated DNA strand of a desired trait and introduce it to a new organism's DNA. And there are two different ways that this is done.

Cisgenic alterations are when DNA is taken from the same species of plant or animal and is introduced into the same species of plant or animal. For example, the DNA that codes for the flavour of an apple is taken and merged with the DNA of the same species of apple in order to increase the flavour. These methods can be used to increase crop yield, increase a plant's hardiness, etc.

Transgenic alterations are when DNA is taken from one species of plant or animal and is introduced into the DNA of another species of plant or animal. One example of this is taking the DNA of luminous jellyfish and introducing it into the DNA of a cat, rabbit, or pig to get a glow-in-the-dark version. One less novel example is taking the DNA of a fish that codes for an anti-freeze protein and introducing it to tomato DNA so that the tomato can withstand colder conditions (Lallanilla, 2016).

Is this procedure safe for human consumption? That is where scientists are currently debating (Freedman, 2013). Some companies have been performing questionable practices in what traits they are crossbreeding, and as a result, you often see their crops pulled before being harvested. Without a single verified case of illness relating to genetic alteration, we can only continue to test GMOs as they are developed (Freedman, 2013). However, you do not need to fear the GMOs that are currently on the market and can be found in your grocery stores. All GMO food that is currently available to consumers is cisgenic (no fish crossed with tomatoes) and has been thoroughly tested. All current transgenic GMOs are still being tested and have yet to be released.

So next time you hear GMO, try not to automatically think of some creature crawling out of a lab, but think of the 10,000 years of history that we have as humans crossbreeding and domesticating plants and animals. Because those are GMOs as well, or just ask the father of genetics, Gregor Mendel, about his GMO experiments (Olby, n.d.).

If you want to learn more about GMOs, be sure to check out this great video from Scientific American's Eric Olson: "What is a Genetically Modified Food?" (A link to the video is listed in the reference appendix).

Thank you for tuning into today's post. And remember, Keep Looking for the Science in Things.

Footnote:

In this blog post, I wrote about GMOs in a very general sense. On the most basic level, a genetically modified organism is an organism whose genetic information/code/DNA has undergone modification through artificial selection. Another great example to think of is dog breeds. Over the thousands of years that humans have called dogs "man's best friend," we have created hundreds of different breed variations by breeding dogs for various different traits (e.g. herding instinct, low shedding, size, colour, etc.). We as humans have genetically modified our canine companions so far from their original prehistoric wolf ancestors, all without a lab.

However, when we often speak of GMOs, especially in agriculture, we very quickly hear the name Monsanto. There are plenty of other companies that produce GMO crop seeds (e.g. DOW Chemical Company, Evogene, and Syngenta), but none of them have been in the media as much as Monsanto. Now, before we dive down the rabbit hole of the "evils" of Monsanto, we need to acknowledge that there were different branches of Monsanto: chemical, agricultural, and biochemical. Founded in 1901, Monsanto Chemical Works started out as a chemical manufacturer that produced saccharin (an artificial sweetener), caffeine, vanillin, and later aspirin (1917) (Britannica, 2023). Troubles for Monsanto started when they became the largest manufacturer of polychlorinated biphenyls (PCBs) from 1935 until 1977 (Neslen, 2017). The manufacturing and sale of PCBs were later banned in 1979 due to their negative impacts on human and environmental health (Nelsen, 2017). Since then, it has become public knowledge that Monsanto was aware of the health and environmental risks of their PCB products, including Aroclor, as far back as 1937, and yet they continued to manufacture and sell them until 8 years later (Nelsen, 2017). In addition to this, as of 2017, they were facing several contamination suits from various city authorities (see Anniston, Alabama, for one of the worst cases) (Nelsen, 2017). But Monsanto didn't stop at PCBs; they later went on to manufacture styrene (1930s/40s), DDT (1944), dioxin (1945), Agent Orange (1960s), RoundUp/glyphosate (1970s), aspartame (1985), and rBGH (Bovine Growth Hormone). I could write an article on how controversial each one of these are and how Monsanto fumbled the ball on most of them, but I will save that for another time. To sum it all up, Monsanto Chemical has an extensive history of questionable practices and using every legal loophole they can find.

And some of those practices have made their way over to the agricultural side. That being said, I couldn't find any studies that could definitively tie Monsanto's GMO seeds themselves to health issues in humans. In researching Monsanto's GMO seeds, I found it to be a bit of a double-edged sword. Many of Monsanto's GMO seeds have been branded "RoundUp Ready." And when most people see that, they immediately think about RoundUp/glyphosate and all its negative health effects. Which they should; however, there is often this assumption that RoundUp, or glyphosate, is in the seed itself. It's not. RoundUp Ready seeds have been genetically modified to be able to withstand the effects of RoundUp/glyphosate, which is a broad-spectrum herbicide. Over the years, many farmers have come to favour using RoundUp, as it is very effective against weeds that would otherwise take over their crops. Many non-Monsanto brand seeds would be wiped out by the herbicide, hence why there is an appeal for farmers to use the RoundUp Ready seeds; less crop weeding for them. So you can probably see where I am going with this. The Monsanto "RoundUp Ready" brand seeds are not necessarily any worse than your garden variety; however, the "RoundUp Ready" crops are dosed and dosed in RoundUp, and that's where the problem actually lies. It is not necessarily the seeds or the crops themselves that are having a negative impact on our health; it is the persistent and toxic chemicals that they are being sprayed with. It would be worth studying Monsanto's "RoundUp Ready" crops without them being treated with RoundUp.

Sidenote: You may have noticed that most articles about Monsanto and their questionable ways seem to have stopped around 2018. Bayer acquired Monsanto in 2018 to the tune of over $60 billion, in addition to being responsible for paying out Monsanto's legal settlements. Time will truly tell if Monsanto and its practices are a thing of the past or if Bayer proves to be worse.

10 Endangered Animals Brought Back From The Brink

(Originally Published May 14, 2018)

Who here has a case of the Mondays? Well, one tried and tested cure that I know of is good news and pictures of some cute baby animals. With Endangered Species Day just around the corner, let's talk about some of the animals that we have brought back from the brink of extinction, and even managed to take off the Endangered Species List.

1. Indian Rhino

As of 1975, there were only 600 individuals left in the wild. After implementing conservation efforts and listing them as a protected species, as of 2012, there are now over 3,500 individuals in the wild, with 645 in Nepal, which prior to being protected no longer had any left ("Greater One-Horned Rhino", n.d.). They are no longer listed as endangered and have been downgraded to vulnerable.

2. Trumpeter Swan

Until the 1980s, they were virtually extinct in the US due to hunting, habitat loss, and lead poisoning. Since being placed on the protected species list, their numbers have made a staggering rebound, and now, as of 2016, the population is around 17,000 in the state of Minnesota alone (Department of Natural Resources Minnesota, 2016).

3. American Bald Eagle

Due to the use of dichloro-diphenyl-trichloroethane (DDT), a pesticide, by the early 1960s, there were only 480 birds in the entirety of the lower 48 states. It progressed to the point that in the 1970s and 1980s, there was only one nesting pair left in New Jersey (Smith et al., n.d.). Fast forward to 2013, thanks to the banning of the use of DDT as well as other conservation efforts, there are 14,000 breeding pairs in the lower 48 states. And as of 2015, there are 161 breeding pairs in New Jersey.

4. West Indian Manatee

As of 1967, only a few hundred were left in Florida's waters, due to habitat changes and disturbances. Thankfully, after being placed on the protected species list, their numbers have increased to over 6,300 in Florida, and around 13,000 worldwide, as of 2016. Thus allowing them to be downgraded from endangered to threatened ("West Indian Manatee," n.d.).

5. Green Sea Turtle

Back in the 1980s, biologists that were monitoring the Green Sea Turtles at Archie Carr National Wildlife Refuge counted less than 50 nests during the breeding season of May 1 to October 1. However, thanks to their efforts, in September 2015, they counted 12,086 nests from the 2015 breeding season (Kotala, 2015). But they still face threats from climate change, habitat destruction, accidental capture in nets, and poaching; thus, they still remain endangered ("Green Turtle: Facts," n.d.).

6. El Segundo Blue Butterfly

Back in 1984, there were a total of 500 individuals on record. But in one of the most amazing comebacks, in 2012, their numbers increased by 20,000%, with over 100,000 individuals being observed out in the wild (Beans, 2013). One place where you may observe them today in great numbers is at Los Angeles International Airport (LAX), where they live in the grasses surrounding the tarmac (Watson, 2016).

7. Southern Sea Otter

Popular for their fur, the Southern Sea Otter was hunted to near extinction, with only 50 individuals known to exist in 1914. Since then, they have faced oil spills, tanker traffic, habitat destruction, and population reduction. In 2016, their numbers exceeded 3,090. In order to no longer be considered endangered under the Endangered Species Act, they must exceed 3,090 three years in a row ("Southern Sea Otter," n.d.).

8. Humpback Whale

Many of us grew up with the slogan, "Save the Whales," and here's why: Back in the mid-1960s, there were only 1,200 humpback whales left in the Northern Pacific due to the whaling industry. Since the whaling industry was banned, they still face threats from pollution, ship strikes, entanglement, and disruption from the offshore oil and gas industries ("Humpback Whale: Facts," n.d.). Today, their numbers sit around 60,000, with many populations now designated as no longer threatened.

9. American Alligator

Another comeback of epic proportions: In the 1950s, the American alligator was near extinction. However, thanks to conservation efforts, as of 2013, their numbers had increased to somewhere over a million individuals out in the wild (The Nature Conservancy, n.d.).

10. Warrnambool Little Penguin

Now, I have left possibly the cutest and one of my favourites for last. The Warrnambool Little Penguins (*Middle Island Warrnambool,* n.d.). In the 1999-2000 breeding season, there were 502 individuals recorded on the island. And their numbers continued to drop due to fox predation. Foxes are among several nuisance species that were introduced to Australia by settlers and have gone on to destroy much of the indigenous animal populations. In 2005, the Warrnambool Penguins' numbers had decreased to less than 10. Desperate to save these little birds, biologists turned to a local chicken farmer, who suggested the use of Maremma guard dogs to watch over the penguins during the breeding season. Introducing Eudy and Tula, two sisters whose current job is to watch over the penguins during the breeding season. Since the introduction of Maremma dogs on the island, there hasn't been a single penguin mortality due to fox predation. As of the 2014-2015 breeding season, there were around 130 little penguins.

Hopefully, that helped with your case of the Mondays. In the meantime, Keep Looking for the Science in Things.

Footnote:

The Indian Rhino population, as of May 2022, sits at 4,014 according to the International Rhino Foundation (Newsroom, 2022).

According to various sources, the Trumpeter Swan population is sitting around 46,000 to 63,000, with numbers steadily increasing every year, so much so that they have been recently classified as being of "Least Concern" on the IUCN Red List ("Trumpeter

Swan,"n.d.) Population studies on Trumpeter swans are conducted every 5 years, and a new study is due shortly.

According to the Federal Register put out in February 2022 by the US Fish and Wildlife Service, the estimated population size of American Bald Eagles in 2019 for the 4 Eagle Management Units (Atlantic, Mississippi, Central, and Pacific North Flyways) is approximated at 316,708 eagles (U.S. Fish and Wildlife Service, 2022).

As of 2022, the West Indian manatee is listed as being "vulnerable." The last aerial count of the Florida population of West Indian Manatees was conducted in 2019, which revealed that there are at least 5,733 manatees in Florida ("Population," n.d.)

The Green Sea turtle population has been on the steady increase with success stories coming out of Cyprus, Costa Rica, and the Seychelles. In the Seychelles alone, an average of 15297 egg clutches (groupings of eggs) have been laid each year from 2014 to 2019 (Pritchard et al., 2022).

No new population studies of the El Segundo Blue Butterfly have been released; however, recently, concerns have been raised over the increased LAX air traffic with LA winning the bid to host the 2028 Summer Olympics and what it will do to the population of butterflies that call LAX home.

The last stock assessment of the Southern Sea Otter was conducted in 2019, which placed the population somewhere between 2,962 and 3,117 (Hatfield et al., 2019).

Although there has not been a global population assessment of the humpback whale, smaller populations have experienced a swell in numbers. In Australia, the humpback whale has been removed from their "Threatened Species List" as a result of almost recovering to their pre-hunted numbers (Long, 2022).

As of 2022, the American alligator is still holding steady in their numbers, despite the "nuisance" alligators being killed in the thousands every year, with over 8,000 being killed in 2021 alone ('Statewide Nuisance Alligator Program, n.d.). Fortunately, the American alligator still remains federally protected by the Endangered Species Act (Florida Fish and Wildlife Conservation Commission, n.d.).

As for the Warrnambool Little Penguins, their guardians Eudy and Tula have since been retired, in 2020-2021 and 2019-2020 seasons, respectively. They are now being guarded by Mezzo, Isola, and Oberon. No evidence of foxes was detected on the island during the 2021/22 season. However, due to poor weather conditions and other extenuating circumstances, volunteers were unable to get an accurate count for the 2021/22 season. The last population estimate was given during the 2019/20 season, with a total estimate of 70 to 100 penguins (*Middle Island Warrnambool*).

Despite the continued success of many of these species, researchers are still watching them closely as certain threats to populations (increasing global and ocean temperatures, continued habitat loss and disruption, lower records of breeding activities being observed, etc.) remain a concern.

Un-Bee-lievable

(Originally Published June 13, 2018)

There's a group of animals that are disappearing at an alarming rate. They are responsible for 70% of the World's food supply and feeding 90% of the Earth's population (Moate, 2014). Without them, the World will cease to exist as we know it. Despite the huge responsibility placed on this animal, it is one of the smallest. In fact, their carrying-ons go almost entirely unnoticed by us. And this animal is the bee.

Honey bees, which are some of the most widely recognized species of bees, live in highly complex community systems, which we refer to as hives. Every member of the community has a role that contributes to the community, whether it is hive construction, gathering pollen, guarding the hive, producing offspring, or caring for the offspring. Not only do they pollinate millions of plants each year, but they also produce liquid gold, which we know as honey.

Bee populations have been in decline since World War II (Spivak, 2013). Over the past few years alone, several species of bees have been placed on the Endangered Species List (Bennet, 2016). And there are a few reasons for this.

Pesticides

Since the conclusion of World War II, agriculture has seen an upswing in the use of herbicides and pesticides to maintain large crop yields (Daftardar, 2015). Bees can be exposed to pesticides in many different ways, whether they are directly sprayed with them when they land on plants, come in contact with the pesticide residue on the plants, exposure to contaminated water, it is introduced into the colony by the bees bringing in contaminated building material and pollen, or unknowingly build nests in contaminated areas (Hopwood et al., 2016). Through the years, humans have used various pesticides that have had varying effects on bees. In recent years, a new pesticide has been introduced to the market to protect plants against sucking and chewing insects (such as aphids and beetles) and has been described as having a low risk to mammals (humans) compared to other insecticides (Hopwood, n.d.). But it has been proven to be extremely detrimental to bees (Hopwood et al., 2016). Neonicotinoids. When bees come into contact with neonicotinoids, it attacks their central nervous system, impairing their memory, activities, and foraging abilities (Hopwood, n.d.). Because of their induced memory problems, many are unable to return to the hive. For those that are able to return to the hive, they introduce neonicotinoids to the rest of the colony, reducing the colony growth rate and the number of queens that are produced (Hopwood, n.d.). In addition, scientists have found that bees exposed to neonicotinoids are more susceptible to pathogens and illnesses (Hopwood, n.d.). This leads us to another factor in the population of bees plummeting.

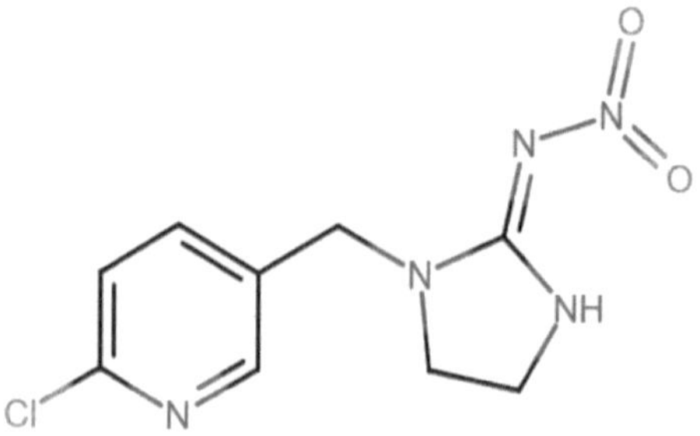

Imidacloprid

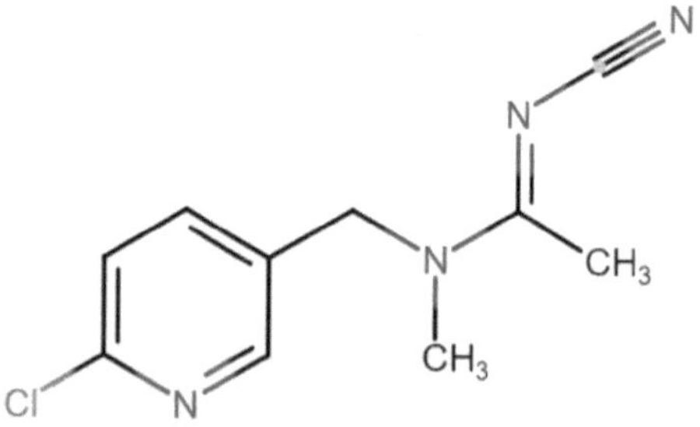

Acetamiprid

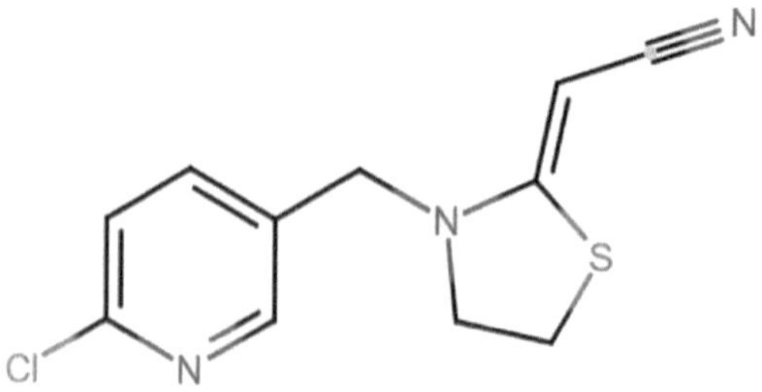

Thiacloprid

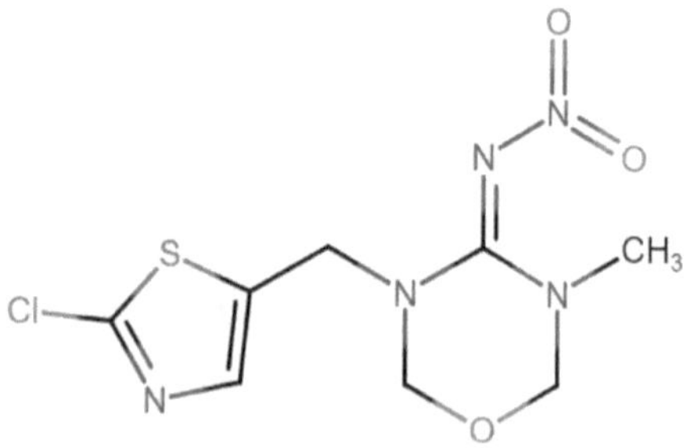

Thiamethoxam

Examples of Some of the Chemicals in the Neonicotinoid Family

Parasites

Like any other living thing, bees do get sick, and they are subject to parasites. The most commonly found parasite is the Varroa mite. This mite relies on the bees and their hives to feed and reproduce. They attach themselves to bees and bee larvae and suck their blood (Bessin, 2016). They weaken adult bees, resulting in them having shorter life spans as they can contract illnesses and viruses easier (Spivak, 2013). As for the larvae, many develop missing wings and legs (Bessin, 2016). Left to their own devices, these mites can destroy entire colonies by killing off the bees. In addition to destroying a colony, they can also be transferred to other colonies via worker and drone bees that travel to other colonies (Bessin, 2016). However, there is treatment available for the colonies that are found to be affected by these bloodsuckers. Colonies that are managed by humans have been salvaged through early detection and treatment (Bessin, 2016). As for the wild colonies, the monitoring and treatment of human-reared bees reduce the spread and impact on the colony.

Decreased Number of Flowering Plants, Increase in Monoculture

Marla Spivak, a leading expert on bees, gave a TED talk in 2013 that highlighted another issue that bees face. In addition to the increased use of herbicides and pesticides since World War II, we, as a society, have also made other changes to our agriculture practices. The number of cover crops that were traditionally used to stabilize soil and replenish its nutrients between crops and to the side of the main crops have been eliminated to make room for primary crops such as wheat, soybeans, corn, etc. These cover crops, such as alfalfa and clover, were excellent sources of pollen and nectar for bees. Another example that Spivak draws attention to is the fact that we have reduced the number of different types of crops that a farmer will produce down to one or two different crops. For her example, she used almond crops. Previously, almond orchards had bee colonies placed among their trees to help pollinate them. Once the trees were harvested, the bees fed on the flowering plants that grew around the orchard and the flowering undergrowth until the trees began to blossom the following year. Now, due to modern almond production, bees can no longer be sustained in the orchards. They must be trucked in for the flowering season and then trucked out; otherwise, they would die.

So What Does Bee Extinction Mean For The World?

If bees disappear entirely, aren't there other pollinators out there that can step in and fill the void? Not exactly. Bees are not the only pollinators out there; there are butterflies, birds, and other insects. Also, humans are able to pollinate plants via paintbrush dusting and other strategies. But bees are the most widespread and numerous pollinators out there. A single hive can contain as many as 60,000 bees, with that hive being capable of visiting around 225,000 flowers each day (Deeley, n.d.). Also, certain plants can only be pollinated by bees. Tomato plant flowers rely on the rumbling of bumblebees to shake loose the pollen (Spivak, 2013). Scientists speculate that we can expect the following foods (among others) to disappear from our grocery stores (Daftardar, 2015):

- Apples
- Asparagus
- Broccoli
- Cacao/Chocolate
- Cashews
- Cherries
- Eggplant
- Kale
- Macadamia nuts
- Nutmeg
- Peaches
- Pumpkins
- Tea
- Almonds
- Blackberries
- Brussel sprouts
- Cantaloupe
- Cauliflower
- Citrus fruits
- Fennel
- Leek
- Mango
- Onion
- Pears
- Raspberries
- Watermelon
- Artichokes
- Blueberries
- Cabbage
- Carrots
- Celery
- Dill
- Garlic
- Lychee
- Mustard
- Passionfruit
- Plums
- Squash

In addition to very slim pickings in the produce aisle, our dairy cooler and meat departments will look rather empty as well. Livestock such as cattle and sheep rely on hay crops such as alfalfa and clover to sustain themselves. Alfalfa and clover receive most of their pollination from, you guessed it, bees. While we are still at the grocery store, how many people cook with canola oil or sunflower oil? No bees, no canola, no sunflowers. And these plants are not only used in cooking oil but in biofuels as well.

Sure, some things do not rely on pollination, such as wheat, rice, and other grasses. And we could adapt by switching from cotton-based fabrics to bamboo-based fabrics. But we would also be placing unnecessary strain on these systems and plants to fill the void that we, as humans, could have helped prevent.

Things You Can Do to Help

Want to help save the bees? Well, you are in luck. We've assembled a list of easy things that you and your family can do from home.

- Purchase plants that haven't been treated with neonicotinoids. Unfortunately, there are some companies out there that sell plants that have been treated with them. Your best bet is going to a local greenhouse, where you can talk with very knowledgeable individuals about their growing practices. Most of the big chains, such as Home Depot, Rona, Walmart, and others, are jumping on the bandwagon with plans to phase out neonicotinoids in the years to come if they haven't already. But there are some who still haven't made the commitment yet. If you're in doubt, a quick Google search of your local major retailer can provide some clarity.

- Plant bee-friendly plants. There are so many plants out there that you could be planting to help the bees. From daffodils, forget-me-nots, pussywillows, rhododendron, and rosemary in the Spring; to sweet peas, fennel, foxglove, thyme, dahlias, heather, lemon balm, and lavender in Summer. A quick internet search will give you hundreds of plants suitable for your planting needs. Whether you want all flowers, edible plants, or plants that are suitable for containers and growing on balconies. Also, if you are in Canada, you can sign up for your free bee-friendly wildflower seed pack through Bees Matter.

- Set up an energy replenishment stand. Bees often run low on energy as they hurry about gathering pollen, and sometimes they require a little rest or even a bit of a pick-me-up. Take a shallow dish (one that you are ok with leaving outside), and make sugary water (use white granulated sugar, NOT brown sugar). Place some rocks in the water so bees have a place to land and hold on to to avoid falling in.

- Put up bee houses. Not all bees build their own hives. Rather, some live in nooks and crannies in trees and logs. Sometimes these spaces are rather temporary, and these bee homes are disturbed and destroyed. But you can make your own bee houses, or if you aren't into the whole DIY thing, you can purchase one of many stylish options.

- Support your local apiaries. These are the people who raise bees and produce honey. In addition, they also help support and grow the bee population in your local area. So whether you buy your honey directly from them, donate, or even spend a day learning from them on how to better help their bees and wild bees, you are helping to boost the bee population in your local area.

In the past several decades, bees have been clinging on as their numbers continue to plummet. On average, around 30% of colonies are lost every Winter (Spivak, 2013). And we, as humans, haven't been a big help. From changing our agriculture practices to spraying pesticides and herbicides that are killing bees, it's about time we stepped up and helped out the little guy. If we all work together, we can prevent losing bees forever. And remember, Keep Looking for the Science in Things.

<u>Footnote:</u>

Unfortunately, bee populations have continued to decline since I wrote the blog post back in 2018. Also, recent reports have stated that our records may not be accurate as they mostly followed domesticated bee colonies and don't account for wild populations, which didn't have the benefit of humans intervening when needed. According to a study by Zattara and Aizen, approximately 25% fewer species of bees were found between 2006 and 2015 than pre-1990, despite an increase in available data.

However, despite continued concerns, some steps in the right direction have been taken:

- Neonicotinoids have been completely banned in the European Union, other than for use in closed greenhouses, as of 2018 (Carrington, 2018).
- In 2019, York University and UBC launched a $10 million, four-year genomic project to research the DNA of bees in search of biomarkers/signs of stress that can be triggered by changing environmental factors (Boisvert, 2019).
- In 2019, actor Morgan Freeman converted his 124-acre ranch in Mississippi into a bee sanctuary (Nace, 2019).
- "No Mow May" was started in 2019 in the United Kingdom as an initiative to promote the growth of natural plants such as clovers, dandelions, and other "weedy" flowers that will provide easy pollen sources for bees and other pollinators as they recover from Winter (Liao, 2022).

<u>They Are Taking Over: Aquatic Invasive Species</u>

(Originally Published July 19, 2018)

It's that time of year again. The weather is hot, and everyone is trying to cool off in any way they can. Those who can head down to the water, do. Everyone with a boat is out on the water, whether they are tubing, waterskiing, fishing, or cruising around. Anyone that has been around boats has been well versed in boat safety: wear a lifejacket; don't operate watercraft under the influence. But one aspect that is less discussed is preventing the spread of invasive aquatic species. Last Wednesday, I was fortunate to attend a workshop about aquatic invasive species and what the Province of Alberta is doing to prevent the introduction and spread of some of these highly aggressive species. And now I'd like to share some of this information with you and how we can prevent its spread.

Aquatic Invasive Species

So what is an aquatic invasive species exactly? Well, they are either a non-native species that has been introduced into an area or a native species that is expanding its range. What makes them a nuisance is that they have little to no natural predators, so they are able to multiply relatively unimpeded. And the reason they are of concern is that they generally pose a risk to the environment, economy, human health, or a combination of these factors.

Now, normally, these species would be relatively contained. They would stay within a local system, as geography would limit the spread (it's hard for an aquatic species to leave the water on its own). However, thanks to humans and our activities, we have helped these species spread. In fact, anything can move around the World in 24 hours nowadays with our rapid trade systems if we aren't careful. In North America, most aquatic invasive species are transported in two main ways: through the transportation of watercraft between bodies of water and through intentional release.

Currently, in Alberta, there are several species that are being monitored and controlled, including Zebra and Quagga Mussels, Crayfish, Goldfish, Phragmites, and Flowering Rush. Each case is unique to each other, and that is why these key examples are the best to learn from.

Zebra & Quagga Mussels

Native to Eurasia, scientists believe that zebra and quagga mussels showed up in North America in the 1980s, most likely brought over in the ballast water of transatlantic vessels ("Zebra and Quagga Mussels," n.d.). And they have quickly become a problem, as one female mussel can produce 1 million eggs each year. They can attach themselves to anything submerged in the water: boats, piping, piers, docks, and rocks. Quagga mussels are even able to anchor themselves to softer materials such as inflatable boats, belly boats, etc. In boats, they can anchor themselves anywhere water is or has been: hulls, bilge pumps, and inside engines. You're probably thinking, "If I take my boat out of the water for a few days, I should be good, right?" Wrong; these mussels are capable of living out of water for up to 30 days.

So they are quite easily transported from one body of water to another. And they can cause quite a bit of havoc once they get there. One mussel is capable of filtering 1 litre of water a day. In most cases, filtering is good, but not in this case. Zebra and quagga mussels filter out all the nutrients, oxygen, and other life-providing aspects of the water and spit out waste. This leads to increased occurrences of blue-green algae blooms, which are toxic if ingested by humans and animals (most blue-green algae advisory warnings usually involve a complete shutdown of the concerning lake – no water access). In addition, the Government of Alberta performed a recent assessment and determined that if Alberta was to become infested with zebra and quagga mussels, there would be an additional $75 million cost every year in order to maintain in-water infrastructure.

Fortunately, as of the latest checks, Alberta does not have any zebra or quagga mussels, thanks in part to preventative measures being taken by the Alberta Government (more on that later on). Unfortunately, that cannot be said for some other noteworthy bodies of water. Lake Winnipeg is a sad case of how destructive these molluscs can be. The species was confirmed to be present in 2013, and by 2016, the lake was completely infested.

Crayfish

Native to many watersheds, crayfish have been seen in other bodies of water where they had not in the past. They are an example of a native species expanding its range, and it's not doing it alone. Scientists have equated this recent spread of populations to being transported by humans to their new homes. Whether it is a case of caught species being released after capture in the closest body of water or a case of someone trying to establish a population that they can later harvest, they have been quickly adapting to their new homes. No assessment has been conducted, so it is yet to be seen what the future impacts of these ecosystems will be. Commonly found in moving rivers and streams, they are quite easily captured (be sure to check with your local fishing regulations if you wish to go and catch some for dinner).

Goldfish

Goldfish - yes, the pet store variety - that you can buy for under $1. Lately, when people find that they can no longer take care of their finned friend (whether their kid no longer looks after it or they are moving away and they can't take it with them), they release it in the nearest storm drain or pond. Unfortunately, many of these ponds drain into natural bodies of water where the goldfish flourish and push out native species. And that finned friend that was the size of a large coin now grows to a whopping half kilogram. In addition to their increase in size, they are able to spawn multiple times a year, releasing 1000 eggs each time, so long as the water is 12 degrees. Native fish only spawn once a year and rely on very specific conditions in order for it to occur, so their offspring are quickly outnumbered and unable to compete. One case of goldfish out of control occurred in a stormwater collection pond in St. Albert, Alberta (just north of Edmonton). By the time the authorities were able to intervene and trawl the pond, they had removed 45,000 individuals. Fortunately, the pond was a closed system, so the goldfish were unable to spread.

Phragmites

Phragmites is a reed that can reach over 4.5 metres. They grow in such dense thickets that animals commonly get stuck and die; they obstruct the flow of water; and they are known to reduce habitat for fish, birds, and other wildlife. Some varieties of phragmites are native, and the only way to tell the difference is through DNA testing. What makes phragmites spread so quickly and far-stretching is that they have two ways of spreading. Like most plants, they have seeds that are easily transported. Commonly found along the railway, it has been speculated that the railway is how it has been spreading. Once rooted, phragmites have an extensive rhizome system, which allows new reeds to pop up from the ground, allowing a single individual plant to cover large areas. Unfortunately, the reed cannot be picked or cut down, as even the smallest fragment can create a new plant.

Flowering Rush

Now illegal in British Columbia and Alberta, this aggressive plant was available for purchase as a pond plant up until 2010. It quickly became a problem as it is capable of out-competing everything. Capable of floating to new places to spread, the smallest fragment of the plant can spawn a plant reaching up to 4.5 meters underwater and creating dense mats of plant mass (solid enough to support the weight of an adult human). However, birds are incapable of nesting in it, and fish can't spawn in it. Back in 2013, when Calgary, Alberta, flooded, a group of plants that had been located in the Calgary Zoo pond was flushed out and swept downstream. Ever since flowering rush has taken over the Bow River into Saskatchewan. And once it reaches a large enough size, it is almost impossible to eradicate. To date, authorities and scientists have tried benthic mats to block out the Sun (it just grows around the mat), mechanical removal with boats (just broke it into smaller pieces, which allowed it to further spread), steam (couldn't get the water hot enough), and scuba divers with vacuums (very expensive, time-consuming, and only mildly successful). The only success they have had so far is in prevention.

What Is Being Done to Prevent the Spread?

Currently, the best tactic for tackling these species is prevention, as for many, once they take over an area, there isn't much that can be done at this point. Both the Canadian and Alberta governments have implemented regulations to prevent the spread. These include the Prohibited Species List, which excludes species that cannot be brought into the country (including the Zebra and Quagga mussels); quarantine provisions for fish being brought in for the pet and aquarium industries to prevent the spread of diseases and parasites; allowing RCMP and peace officers to be able to enforce the Aquatic Invasive Species Act; prohibiting the release of subject water (including aquarium water) and non-native species, among other mandates. In addition, scientists and conservation officials are currently monitoring 112 locations throughout Alberta alone to see if there are any new invasive species introduced or if existing populations do not spread further.

 In order to ensure that species do not enter Alberta and Canada unknowingly, there are mandatory watercraft inspection points set up at various points. In Alberta, as well as other locations, it is mandatory to pull into these inspection stops if you have a watercraft of any kind, trailered or inflatable. And at some of these locations, conservation officers are assisted in their inspection by very special dogs. It is important to note that these dogs are strictly trained to find invasive species (in the case of boat inspections, they are looking for zebra and quagga mussels); they are not looking for drugs, alcohol, or firearms. As much as they can be a concern, they are not a concern for these dogs and their handlers. Last Wednesday, I had the opportunity to meet one of these working dogs, Seuss, along with his handler. Seuss and the other two dogs that work in Alberta, Hilo and Diesel, come from the program

Working Dogs for Conservation, based in Montana. Dogs that are a part of this program are rescued from shelters and are trained to fill various positions, from finding rare animals to searching for invasive species and detecting smuggled or poached animals. In the case of Seuss, Hilo, and Diesel, they are trained to search for zebra mussels (in boat inspections, shoreline checks, and water samples), Thesium plants, and are currently learning how to detect wild boars via their scat. Without these dogs, inspections would take much longer, and there is a chance that some could be missed by the human eye, which can be picked up by one of the dogs' keen sense of smell.

What Can You Do?

Want to help prevent the spread of aquatic invasive species but don't know how you can contribute? Well, you are in luck; there are plenty of ways:

- Clean, Drain, Dry Your Boat and Gear – especially when going between different bodies of water. Not only does it prevent stowaway species, but it also prevents the spread of diseases and parasites that could negatively impact an ecosystem. Be sure to pull your ballast plug, it is mandatory and prevents the spread of invasive species, diseases, and parasites. As for drying your gear, the best practice is to allow it to dry outside in the Sun for 24 hours.

- Don't Flush Your Goldfish – flushing your dead goldfish could potentially introduce bacteria or viruses into the local water system. For dead pet fish, it is recommended that you bury them;

there are even biodegradable burial pods available, like Fishpods by Pawpods. If you have a pet fish that you don't want any more and no one will take it, whatever you do, do not release it into the wild, as it can quickly overrun a water system. One more humane method to euthanize your fish is by putting them in a plastic bag with water and placing them in the freezer. As the water cools, the fish slows down and drifts off to the other side, so to speak.

- Don't Let It Loose – In addition to your pet goldfish, do not release other species or live bait into different bodies of water from which they were caught or used.

- Report It – If you spot an invasive species, report it. Even if you are in doubt, report it. I spoke with one of the individuals who sees some of the reports, and she said that she would rather have a report that turns out to be a native plant that is not a concern than have an invasive species go unreported. There are two different ways to report. In Alberta, there is a 24/7 hotline that can be called at 1-855-336-BOAT (2628). When you call in, be sure to report the date and time; the river, stream, or lake name; the GPS coordinate (if you have it); the description (number of invasive species individuals, species, etc.); and take photos if possible. A second way to report, which is being introduced in Alberta and is currently being used in various places throughout North America, is the app EDDMapS. Through the app, you submit a picture of the invasive species, the app takes the location from your phone, and it develops a report for you.

- Don't Pull or Dig It – Any invasive plants that you see, don't pull them out or dig them up. Many species only need a small piece to create a new plant. Also, some species of invasive plants can be harmful to you. Rather than trying to remove it, report it so the proper authorities can remove it properly, preventing it from spreading further.

Aquatic invasive species are a serious issue for our environment, our economy, and our health. It is an issue that is not going to go away with our help and due diligence. Many invasive species still have no tried-and-true method to eradicate them, so prevention is the best means we have right now. If we all help out, we can prevent the destruction of more ecosystems. For more information, check your local fishing and boating regulations. In Alberta, you can get further information at aep.alberta.ca. That's all from us today; we have much more information from the workshop, but that will be another post. And remember to Keep Looking for the Science in Things.

Footnote:

Here are a few updates since I wrote this post in 2018:

- As of 2020 in North America, zebra mussels have been found in Lake St. Clair, the Great Lakes, the St. Lawrence River, the Mississippi watershed, Lake Winnipeg, the Red River, and the Nelson River (DFO, 2021).

- As of 2020 in Canada, Quagga mussels have so far been limited to the Southern Great Lakes (Ontario, Michigan, Huron, and Erie), St. Lawrence River up to Quebec City (DFO, 2020).

- As of September 2020, the flowering rush is still proving to be a persistent issue in the Bow River and has fully infested the South Saskatchewan River in Alberta and leading into Saskatchewan (see the link to Flowering Rush Distribution Map in Appendix).

- As of writing this footnote (June 2023), Hilo and Diesel are still listed by Working Dogs for Conservation as still working. Both Hilo and Diesel have learned to consistently identify three scents each.

Diving Into the World of Sharks

(Originally Published July 24, 2018)

It is that time of year again that has all of us Selachiiphiles chomping at the bit. If you haven't already guessed it, it's the Discovery Channel's Shark Week. And not just any shark week, it is the 30th anniversary of Shark Week. 30 years of watching the silent predator of the deep from the comfort of your living room. But what makes us so fascinated with this apex predator that has outlived even the dinosaurs? Well, in today's post, we are going to dive into the world of sharks and find out the many reasons why we find them so captivating (hopefully without too many puns).

What Are Sharks?

Sharks belong to a family of fish that includes rays and skates, which are unique in their own right ("Shark: Facts," n.d.). The fish that make up this group can be identified by three distinguishing features: five- to seven-gill slits, a skeleton made up of cartilage, and pectoral fins that are not fused to their heads. Sharks are known to live between 20 and 30 years in the wild, although some older individuals have been recorded to live up to 100 years old ("Sharks and Rays," n.d.). There are over 500 different species of sharks in existence that we know of. With more species being discovered each year. Ranging in size, they vary from the length of a human hand to being more than 12 metres long (The

Ocean Portal Team, n.d.). Not only do they vary in size, but in habitat as well: from the frigid waters of the Arctic to tropical reefs to the open ocean and at great depths (The Ocean Portal Team, n.d.). Many have called these animals living dinosaurs. However, these predators have been swimming in the ocean's waters for 425 to 455 million years (Martin, n.d.). Unlike many other fish, some shark species lay eggs, while other species give birth to live young (much like marine mammals) (University of Florida, n.d.). Of those that give birth to live young, it is not uncommon to see a mother shark accompanied by her pups as she hunts. As for food, their diet mainly depends on their size, but it mostly consists of fish, crustaceans, molluscs, plankton, krill, marine mammals, and other sharks. In fact, the largest shark and my personal favourite, the whale shark, measuring over 12 meters in length, sustains itself by filter feeding on plankton ("Whale Shark," n.d.).

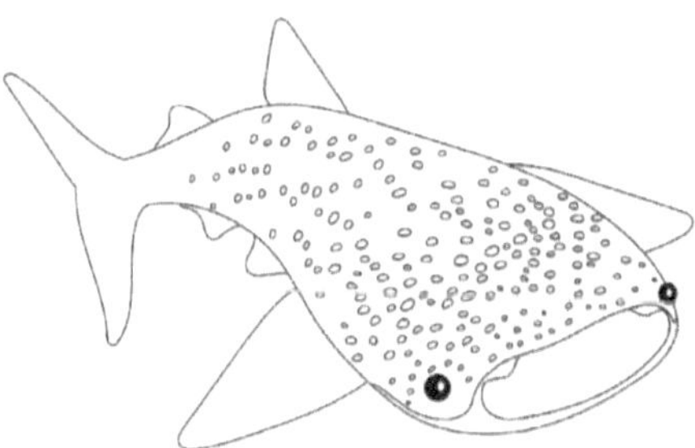

Sharks Through the Ages

Despite the vast variety of sharks present today, there are many others that have graced the World's oceans but aren't present today. Early sharks are speculated to have either been freshwater dwellers or stuck close to the coast (The Ocean Portal Team, n.d.). But other than that, very little information is known about them. About 370 million years ago, sharks became closer to what they are today. They are believed to be around 1.22 metres in length, with their mouth at the front of their head rather than under it (like modern-day sharks), and they were known to dwell in both fresh and saltwater (The Ocean Portal Team, n.d.). From 360 to 286 million years ago was the Golden Age of Sharks, which resulted in a wide array of adaptations and characteristics to appear (The Ocean Portal Team, n.d.). From sword-like appendages on their heads, to head spines, to anvil-shaped dorsal fins, to buzzsaw-like lower jaws (The Ocean Portal Team, n.d.). Unfortunately, many of these species were wiped out, along with the rest of the 95% of animals on Earth, during the Permian-Triassic Extinction, 251 million years ago. But those sharks remained and began their climb to the position of apex predator (The Ocean Portal Team, n.d.). Increasing in size to between 6 and 10 meters, these species competed alongside plesiosaurs and mosasaurs for food and territory (The Ocean Portal Team, n.d.), until 65 million years ago, when the Cretaceous-Paleogene Extinction occurred, which is best known for wiping out the dinosaurs (The Ocean Portal Team, n.d.). But not sharks, with about 80% of their species surviving, many of which are ancestors to today's sharks (The Ocean Portal Team, n.d.). With this large void in the ocean's ecosystems, sharks were able to flourish and become the predators that they are today, as well as some grew to super-sizes, e.g. Megalodon, which existed from 16 million years to 1.6 million years ago (The Ocean Portal Team, n.d.).

Jaws: Sharky Hits Hollywood

Today, it would be very difficult to find someone who hasn't heard of the infamous Jaws. The blockbuster hit terrified the World with its portrayal of the bloodthirsty, man-eating great white shark that preyed on beachgoers. It can also be accredited to humans' distrust of sharks and our need to hunt them down, causing their numbers to plummet. But in actual fact, you are more likely to die from drowning at the beach (chances of 1 in 3.5 million) than being attacked (1 in 11.5 million) and killed (0 in 264.1 million) by a shark (University of Florida, n.d.). So why has Jaws terrorized us out of the water? Well, that's because it actually happened, well sort of. The inspiration for the novel that later led to the movie Jaws comes from a series of shark attacks in July 1916, which left four dead (including the first fatal shark attack in US history) and one injured over a span of almost two weeks (Capuzzo, 2016). This hunter is speculated to be a young great white that was patrolling the

Jersey Shore looking for an easy meal (Capuzzo, 2016). And the similarities between the movie and the real event don't end there: one victim was attacked in an estuary, and another died while trying to wrestle with the shark in an attempt to kill the beast (Capuzzo, 2016). It is because of this dramatization of history that many of us can be chased out of the ocean by someone singing two ominous notes, dun-dun (including my younger self, thanks Dad). But it's important to remember there have only been around 4,120 unprovoked attacks ever recorded, and of those, only 1,013 have been fatal, whereas, in 2013, 1.25 million people died in traffic accidents ("Incidents by country, n.d. | WHO, n.d.). You're more likely to die slipping in the bathroom after you get out of the shower than being attacked by a shark.

Species At Risk

Over the past 425 to 455 million years, sharks have patrolled the ocean's waters with little to no threat from other species. Unfortunately, due to Hollywood's portrayal, our fear of the unknown, the fishing industry, and their "luxuryesque" appeal, sharks' numbers have been on a rapid decline. It is speculated that 100 million sharks are killed by the commercial and recreational fishing industries alone (The Ocean Portal Team, n.d.). Many of which are caught just for their dorsal fin, which is the main ingredient in the "luxurious" status symbol that is shark fin soup, which drives a trade industry that handles between 100 and 273 million fins each year (The Ocean Portal Team, n.d.). Even if they are not caught intentionally, many shark species are threatened as a result of

by-catch, and many do not survive being caught and thrown
back. Almost half of all pelagic sharks are threatened, and 25% of all
sharks and rays that inhabit coastal regions are threatened, with another
35% unknown ("Shark: Facts," n.d.). And this has had a profound
impact on the ocean's ecosystems, many of which have yet to be seen.

Sharks have long been portrayed as bloodthirsty monsters of the deep.
But they are also survivors who have silently glided through the Earth's
oceans for over 400 million years. They are nurturing parents who look
after their young after they are born. And they have been overseers of
the ocean's ecosystem and have maintained a delicate balance since
before the time of the dinosaurs. If we are careful, their legacy will die at
our hands. So rather than fear sharks, I encourage you to learn more,
tune into Shark Week, and check them out at your local aquarium. They
might just amaze you, like they did me. And you don't have to dive into
a cage among great whites. I got to pet and feed bamboo sharks at a local
aquarium, and it was pretty cool. I hope to one day get to swim with
them. That's all we've got for today. In the meantime, Keep Looking For
the Science in Things.

Left-Handed, Right-Handed

(Originally Published August 14, 2018)

There are two types of people in this World. Left-handed people and right-handed people. Yesterday marked International Left-Handed Day. But what determines whether a person ends up being right-handed or left-handed? Is it genetics, an acquired trait, or something else? In today's post, we will be looking into why you ended up either left-handed or right-handed.

Born This Way

According to recent studies, the underlying factor that determines whether your dominant hand is your right or left is your genetics (Klar, 2003). However, it's not a single gene or set of genes, like those that would code for other traits such as hair colour, eye colour, gender, etc., but rather a "network" of genes that code for whether you use your right or left hand. So, whether you are right-handed or left-handed, you have your parents to thank for it, including the annoying fact that you may smudge your writing. Despite this major determining factor, there are a few other factors that can further influence an individual's dominant hand.

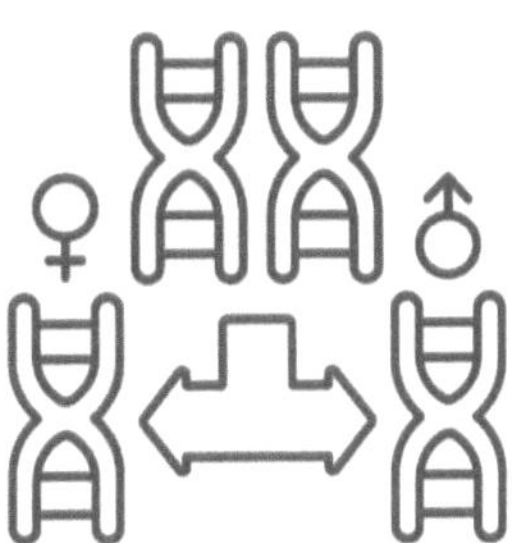

Where Do You Interpret Language?

Another study indicates that your handedness could be related to the structure of your brain, in particular, your motor cortex and the areas of the brain that are associated with language (Brandler et al., 2013). According to Knecht et al., if you analyze language primarily on the right side of your brain, you are more likely to be left-handed, and if you analyze language primarily on the left side of your brain, you are more likely to be right-handed. This could possibly be why artists and other creatives may have a higher prevalence of those that are left-handed than those that are more analytical (something that may be worth looking into). This speculation is due to the fact that creativity is processed by the right side of the brain, which controls the left side of the body, and analytical thinking is processed by the left side of the brain, which controls the right side of the body ("Left and Right Hemisphere, n.d.). What about those that analyze language 50/50 with the left and right sides of the brain? Well, the average human brain processes language asymmetrically, so it is either processed more on the left side or the right side of the brain; there is no even split. That being said, scientists have even noted that the less asymmetricality the brain exhibits, the more other things arise that affect brain structure and chemistry. Schizophrenia has been linked to reduced asymmetry in a section of the language area of the brain, the planum temporale. In addition to this, those who are schizophrenic, due to this reduction in asymmetry, have a higher likelihood of being left-handed, even if they are genetically more likely to be right-handed (Sommer et al., 2001 | Dragovic and Hammond, 2005).

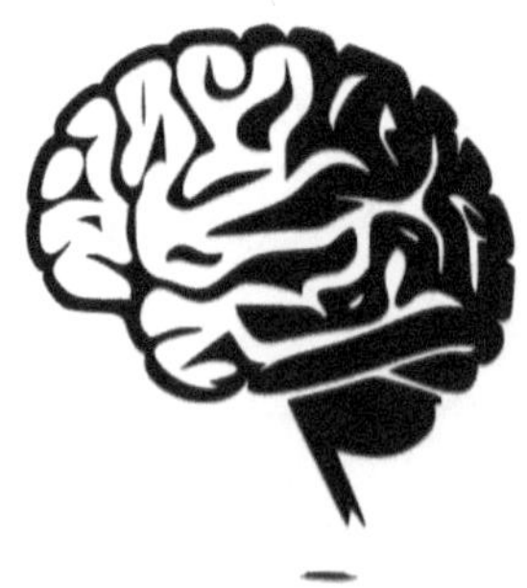

Left-Handed Witch Hunt

What about the countries that have no left-handed people? Is that a genetic phenomenon or something more systemic? Countries such as China, India, and certain countries in North Africa, East Africa, and the Middle East have very few people who are left-handed, so it could be considered statistically impossible. This is because being left-handed is culturally or religiously frowned upon in these countries. In India, traditionally, your left hand is used in the bathroom, so to use that hand outside of the bathroom is considered unclean (Cook, n.d.). Some countries go as far as condemning the left-handed individual as evil, a witch, or being possessed by the Devil (Geilig, 2013). As a result of this negative attitude towards being left-handed, there is a system put into place to 'cure' these afflicted people (Kushner, 2013). In fact, it wasn't very long ago that other countries, including Canada, punished children in school who wrote with their left hand in hopes of correcting this "problem". As a result of these implemented policies and practices, people in these countries become right-handed to avoid being criticized and subjected to "corrective" measures.

Handedness is Not Just for Humans

While researching for this post, we discovered that humans are not the only animals in the animal kingdom that exhibit a preferred or dominant hand. In fact, there are a large variety of animals that have a preferred hand. These include chimpanzees and other primates, rats, cats, dogs, birds, whales, polar bears, and horses, among many others (Finch, 1941 | Yoshioka, 1930 | Cole, 1955 | Tan, 1987 | Mastin, n.d.). In fact, some of these animals even showed evidence of the asymmetry in the brain, similar to humans, that contributes to our handedness.

So whether your dominant hand is your left or your right, it is not due to one sole factor but rather a combination of genetics, brain structure, and culture. Regardless of our dominant hand, it does not dictate whether you are a good person or a bad person; you may just smudge your writing, and scissors may be the bane of your existence. That's all from us today, and remember to Keep Looking for the Science in Things.

Turkey Coma: Fact or Scapegoat?

(Originally Published October 2, 2018)

It's that time of year again, at least in Canada. Next Monday marks Thanksgiving, and as many look forward to the upcoming Monday off, spending the day off, and even the delicious feast, we, here at Schrodinger's Cat, want to set the record straight with the well-known turkey coma. That's right; there's no such thing! What?!?! Check out today's post to learn more.

L-Tryptophan

We have all been taught that after a giant Thanksgiving feast, when we are full to the point of bursting, the sleepy feeling creeping over us is due to a compound found in turkey, L-tryptophan. But what is L-tryptophan? Tryptophan is an essential amino acid that our body uses to make Vitamin B3, which aids in digestion and keeps our skin and nerves healthy (Zamosky, 2009). It's found in not only turkey but other poultry, meat, cheese, yoghurt, fish, and eggs. In addition to all the benefits that tryptophan has, it is also necessary for the production of serotonin, or the "feel good" hormone in our brain that makes us feel relaxed (Ballantyne, 2007). Once serotonin is produced, it is then used by our bodies to produce melatonin, which helps us feel sleepy (Ballantyne, 2007).

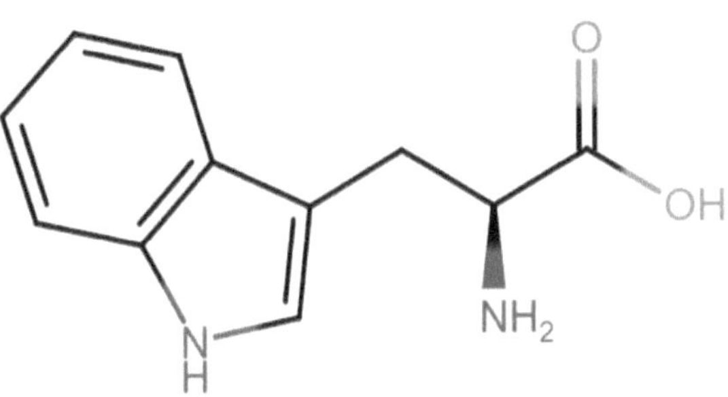

L-Tryptophan

Prime Suspect: Turkey

So, we know that turkey contains tryptophan; doesn't that seem a bit like a smoking gun pointing to the guilty culprit for the post-Thanksgiving coma? But there is a catch. Turkey contains no more tryptophan than any other poultry (Zamosky, 2009). In fact, chicken is known to contain slightly more tryptophan. And yet, we don't pass out on the couch after an evening of fried chicken - at least most of us don't. So if it's not turkey, why is napping the most popular activity on Thanksgiving?

The Secret is in the Combination

The truth about the Thanksgiving sleepiness lies in what else we eat with our turkey. Carbohydrates help tryptophan in its ability to produce serotonin more efficiently (Zamosky, 2009). And what Thanksgiving dinner would be complete without potatoes, buns, corn, stuffing, and dessert? In fact, according to a recent study, if you wish to improve your sleep, a small snack containing around 30 grams of carbs will actually help you sleep better (Zamosky, 2009).

Sugar Crash

In addition to the carbs, you are also certain to enjoy some dessert, which increases the amount of insulin in our bloodstream. Insulin is a hormone that allows the uptake of glucose (sugar) and amino acids (like tryptophan) into our tissues (Ballantyne, 2007). Although insulin is known to have little effect on tryptophan, it uptakes other amino acids from the blood, which allows tryptophan to cross the blood-brain barrier and ensures it begins producing serotonin unimpeded, having maximum sleepy effects (Ballantyne, 2007).

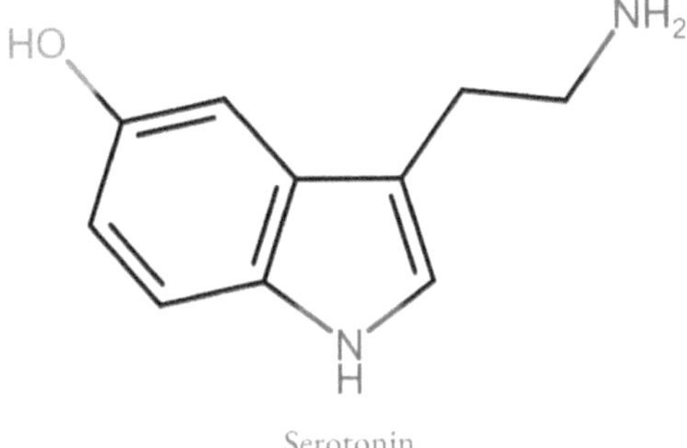

Serotonin

Moderation Is Key

Who here is guilty of overeating at Thanksgiving? I know I definitely do! But it's also part of why we feel so tired (Zamosky, 2009). Our body burns energy when we digest food, and the more food our body has to digest, the more energy we consume. When we normally only eat a plate of food for dinner, we generally don't need a nap. But when we have a heaped plate of food and then go back for seconds of our favourites, and finally top it off with dessert, which includes a piece of each pie, because why should we have to choose? And before we know it, we are on the couch, snoring.

So, if you wish to ditch the turkey coma after your Thanksgiving dinner, feel free to enjoy the turkey; just watch your carbs and sugars, and of course, exercise portion control. But really, what's so bad about a Thanksgiving nap? After all, it's a holiday, not every day. So heap up the plate, taste all the delicious food, and enjoy your nap. Just don't frame the turkey. From us at Schrodinger's Cat, we wish everyone a Happy Thanksgiving. If you are travelling to see family during the holiday weekend, we wish you safe travels. In the meantime, Keep Looking for The Science in Things.

<u>The Great Avian Migration</u>

(Originally Published October 16, 2018)

You know it's Autumn when the birds take to the sky and begin their migration south. Sure, the older birds have done the route before, and they may guide the younger ones. But how did they figure it out in the beginning, and what about those that don't have any veteran birds to guide the way? There is no one navigation strategy that is used by birds in order to fly south and then fly home again. Some species may use one method, while others use an entirely different strategy. So, in today's post, we will be talking about the different ways birds conquer the migration of thousands of kilometres.

"Third Star On The Right, And Straight On Til Morning"

Like many animals, including humans, birds use landmarks in order to navigate during their migration (Connor, 2008). Whether it is a stationary landmark, such as a tree, lake, or building, or whether it is more transient, like the Sun, the Moon, and the stars, birds use the sight and positioning of these landmarks to guide them along their way. Wherever you are on Earth, the Sun, and the Moon rise in the East and set and the West and the North Star is in the northern sky (even though they do move as the day or even the year progresses).

Smells like Home

Smell is another way that birds can tell if they are on the right path or if they have made it to their final destination (Connot, 2008). Certain areas can have a distinct smell, and it really makes sense. Some areas have certain types of plants or flowers growing there, while coastal regions will smell of saltwater, and bogs can have a musty smell. Even humans have olfactory memories. Ever smelled freshly baked bread and been instantly transported back to your grandmother's kitchen? Or smelled a certain floral perfume and thought of your mother sitting at her dressing table getting ready for an evening out with the girls? These olfactory memories are created because the structures of our brains that process smells and store memories are so closely situated (Mouly & Sullivan, 2010). The brain of a bird is not unlike a human brain, at least when it comes to tying scents to memories.

Magnetic Pull

As we all know, the Earth has a naturally occurring magnetic field that humans use, in conjunction with compasses, in order to navigate via the magnetic North Pole. And humans are not the only ones that use this magnetic field to navigate; but other mammals, sea turtles, insects, and even birds do. Some use it much like humans do, using it to point themselves in a certain direction, while others use it more like GPS to determine their exact position (Connor, 2008). But birds don't have compasses or the ability to hold and orientate a compass (it would be pretty hard to flap their wings and hold a compass), so they use an iron molecule known as magnetite that is located in their beak to determine north and south during their migration (Connor, 2008).

A New Theory

In recent studies, scientists have discovered some photo-sensitive molecules (those that are sensitive to light or react to light) that can turn into one of two chemical states based on the direction of the bird in comparison to the Earth's magnetic field (Connor, 2008). This chemical has been found in the retina of a bird's eye, which could allow them to orient themselves at dusk as they continue on their migration through the night (Connor, 2008).

Birds are some of the best navigators in the animal kingdom, and they have many tools that prevent them from becoming disoriented and lost. In the animal kingdom, the difference between being lost and knowing and finding their way is the difference between survival and perishing. The various ways that birds navigate are just a few of the strategies that are used by animals around the World. In the meantime, Keep Looking for the Science in Things.

It Came From The Deep: The Headless Chicken Monster

(Originally Published October 23, 2018)

It's not that often these days that you hear of a new sighting of an animal. But a group of scientists have discovered an animal off the coast of Antarctica that has only ever been seen in the tropical waters of the Gulf of Mexico. Not only is it rather bizarre-looking, but it has been nicknamed the "headless chicken monster". In today's post, we will be discussing what this "monster" really is and what this discovery means for the coastal waters of Antarctica.

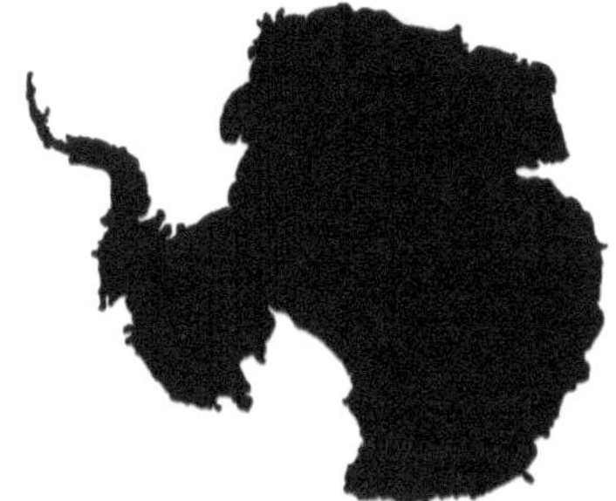

Discovered by a group of Australian scientists while they were monitoring the impacts of commercial fishing on the Antarctic marine ecosystems, the "headless chicken monster" is a deep-sea sea cucumber that has only previously been seen in the Gulf of Mexico (a drastically different ecosystem). Also known as *Enypniastes eximia*, or the Spanish Dancer, this individual was spotted deep in the Southern Ocean, off East Antarctica.

Sea cucumbers are invertebrates belonging to the group of animals called Echinoderms, which include sea urchins, sea stars, brittle stars, and sea lilies (Miller & Pawson, n.d.). A rather slow-moving animal, the sea cucumber escapes predators by spitting out its internal organs, allowing the predator to have an easy meal while the sea cucumber makes it escape ("Sea cucumber," n.d.). But have no fear; the sea cucumber is able to fully regrow its ejected internal organs.

The observation of the "headless chicken monster" was made during a scheduled research expedition on the effects of commercial fishing on marine ecosystems, in particular the Southern Ocean. In addition to this sighting, the scientists have gained a better understanding of the areas that are able to better withstand commercial fishing and the areas that are more sensitive and should be protected. And this sighting serves as a great reminder of our need to protect and manage these ecosystems to ensure their continued survival.

With how little of the ocean has been explored by humans, there are lots of animals and ecosystems that may be disappearing without us even knowing. As humans, we need to pay attention to our actions and the consequences that they cause, as they may lead to the elimination of an entire habitat. If you are looking to make positive choices when it comes to the fish you enjoy at dinner, be sure to look for the Ocean Wise label and avoid fish caught via more destructive practices (long line, bottom trawling, cyanide fishing). In the meantime, Keep Looking for the Science in Things.

Earth + Environmental Sciences

4 Ways to Make the Planet Greener, While Saving You Some Green

(Originally Published April 16, 2018)

Every day, we rely on our planet to provide for us. Whether it be food, shelter, a means of employment, or even just a rock to sit on while we zip through the universe, it could be argued that we need Earth more than it needs us. So why aren't we looking after it a little bit better? Earth Day is just around the corner on Sunday, April 22, and it's the time of year that many of us get out and pick up garbage in our neighbourhood or plant a tree. But what if I told you there were ways to make Earth a greener place while saving you some green (money). How?

Green living and green technology are marketed in such a way that it seems like a luxury that is unattainable for many. Not true. In fact, I have four, yes four, ways that you can help the planet out without hurting your wallet.

1. The Different Types of Lightbulbs You Use

One of the first ways to make Earth a greener place while saving you some green comes down to the lighting in your house. We are all taught from a very young age, "Turn off the light when you leave the room." So that's a given, but have you put much thought into the type of light bulb that is in your house? With so many options in the lighting aisle at the hardware store, you can easily spend a good hour comparing the traditional incandescents to the "green" CFLs and the newer LEDs that we see slowly invading the shelves. So which one is better for Earth, and your wallet? Let's look at the numbers.

The average Incandescent Lightbulb is 60 W.

Average Watt consumption of an Incandescent Lightbulb = 10 – 17 lumen per watt

Total Lumens of Incandescent lightbulb
= (13.3 lumen / watt) x 60 watt = 800 lumens

**which is the average household value*

For a CFL, which has an average watt consumption of 50 to 60 lumen per watt

Total Watts of CFL = 800 lumen x (watt / 55 lumen)
= 14.5 W

LEDs with 800 lumen is about 10 W (as per a box of NOMA LED 60 W replacement lightbulbs)

In many places, you can choose your energy rate. The current average energy rate in Alberta, Canada (as of April 2018) is about $0.0614/kWh. The average person works away from home for 8 hours a day and sleeps for 8 hours a day. Assuming that no lights are on when you are away from the house and while sleeping, that leaves 8 hours in a day that the lights are on.

To calculate kWh,
Energy = (Power x Time) / 1000

For the Incandescent,
E = (60 W x 8 hr/day) / 1000 = 0.48 kWh/day

For the CFL,
E = (14.5 W x 8 hr/day) / 1000 = 0.116 kWh/day

For LEDs,
E = (10 W x 8 hr/day) / 1000 = 0.08 kWh/day

Now that we have the amount of kilowatt-hours that are used by each type of light bulb in a day, we can determine the cost.

For Incandescent,

> *$ per lightbulb per year*
> *= ($0.0614 / kWh) x (0.48 kWh / day) x (365 days / year)*
> *= $10.76/lightbulb/year*

For CFL,

> *$ per lightbulb per year*
> *= ($0.0614 / kWh) x (0.116 kWh / day) x (365 days / year)*
> *= $2.60/lightbulb/year*

For LED,

> *$ per lightbulb per year*
> *= ($0.0614 / kWh) x (0.08 kWh / day) x (365 days / year)*
> *= $1.79/lightbulb/year*

The average 3 bedroom, 2 bathroom house has around 40 light bulbs (a lot of light fixtures have 2+ lightbulbs per fixture, and many homes have lamps in addition to ceiling-mounted fixtures). So the yearly household breakdown on lighting alone (with 40 light bulbs) is around:

> *$398.12 for a house lit with incandescent light bulbs*

> *$ 96.20 for a house lit with CFL light bulbs*

> *$ 66.23 for a house lit with LED light bulbs*

 Now, a lot of people see only the thirty-dollar difference between CFLs and LEDs and go, "Why would I spend a significantly greater amount of money on LEDs when in a year I'm only saving an additional $30?"

There are two reasons to consider LEDs over CFLs:

> 1. One LED light bulb has an average lifespan of 20,000 hours (depending on the brand) (Energy Saver, n.d.). One CFL light bulb has an average life of 6,000 hours (depending on the

brand). Therefore, you will have to replace a CFL more than three times in a single LED's lifespan. So with CFLs costing about $3.62 per light bulb, if you multiply that by three changes, you will spend about $10.87 on CFLs when you could have spent $3.33 on a single LED. So in the same 20,000 hours, in a house with 40 light bulbs, you would spend $434.80 on CFLs or $133.20 on LEDs.

2. Means of disposal. Once your light bulb has burned out, you are obviously going to want to get rid of it. LEDs can be easily recycled at designated recycling depots. However, if you cannot find a nearby recycling depot, in a pinch (and as a last resort), you can always throw LEDs in your household garbage, as there are no harmful toxins in them (Taylor, n.d.). CFLs, on the other hand, are much trickier to dispose of. They should be recycled, not thrown out, as they do contain trace amounts of mercury ("Learn about CFLs, n.d.). Also, if they do break, they can be quite hazardous to yourself, your household, and the environment.

2. Using Your Oven Less

According to Toronto Hydro, the average stove/oven combination is 5000 W and used for about 100 hours each month. So, the average kWh of a stove/oven combo and resulting bills are...

Average kWh each month
= (5000 W x 100 hrs) / 1000
= 500 kWh/month

Average monthly & yearly energy bill for a Stove/Oven combo

Average monthly $
= ($0.0614 / kWh) x (500 kWh / month)
= $30.70/month

Average yearly $
= ($30.70 / month) x (12 months / year)
= $368.40/year

Now, the average microwave oven is around 1000 W and used for about 5 hours each month.

Average kWh each month
= (1000 W x 5 hrs) / 1000
= 5 kWh/month

Average monthly & yearly energy bill for a Microwave Oven

Average monthly $
= ($0.0614 / kWh) x (5 kWh / month)
= $0.31/month

Average yearly $
= ($0.31 / month) x (12 months / year)
= $3.72/year

Finally, another alternative to the conventional oven is the average toaster oven, which is 1500 W and used for about 2 hours each month.

Average kWh each month
= (1500 W x 2 hrs) / 1000
= 3 kWh/month

Average monthly $
= ($0.0614 / kWh) x (3 kWh / month)
= $0.18/month

Average yearly $
= ($0.18 / month) x (12 months / year)
= $2.16/year

So in a year, the average household consumes 508 kWh of energy to power their stove/oven, toaster oven, and microwave, to the tune of $374.28 per year.

Now, what if we used our microwaves and toaster ovens more and our oven less? Obviously, some things must still be cooked in the oven due to size, and it is not recommended to cook raw meat in the microwave. But what if we cut the hours we use the oven in a month in half?

The Stove/Oven still is 5000 W, but now only 50 hours/month.

Average kWh each month
= (5000 W x 50 hrs) / 1000
= 250 kWh/month
Average monthly & yearly energy bill for Stove/Oven

Average monthly $
= ($0.0614 / kWh) x (250 kWh / month)
= $15.35/month

Average yearly $
= ($30.70 / month) x (12 months / year)
= $184.20/year

Next is the microwave, which is still 1000 W. If we take a quarter of the time that the oven is used, that is 25 oven hours. I say oven hours because the time cooked in the microwave is less than in the oven. So in order to convert from oven hours to microwave hours, we must divide the oven hours by 4.

Microwave Oven hours per month (plus the 5 hours we already use it each month)
=(25 hours / 4) + 5 hours
=11.25 hours/month

Average kWh each month
= (1000 W x 11.25 hrs) / 1000
= 11.25kWh/month

Average monthly & yearly energy bill for Microwave Oven

Average monthly $
= ($0.0614 / kWh) x (11.25 kWh / month)
= $0.69/month

Average yearly $
= ($0.69 / month) x (12 months / year)
= $8.28/year

Finally, the toaster oven is still 1500 W and is used for 25 hours of the oven's time, as well as the 2 hours already used each month, for a total of 27 hours per month.

Average kWh each month
= (1500 W x 27 hrs) / 1000
= 40.5 kWh/month

Average monthly & yearly energy bill for Toaster Oven

Average monthly $
= ($0.0614 / kWh) x (40.5 kWh / month)
= $2.49/month

Average yearly $
= ($2.49 / month) x (12 months / year)
= $29.88/year

So, with our altered usage in a year, the average household would consume about 302 kWh of energy to power their stove/oven, toaster oven, and microwave, to the tune of $222.36 per year. Saving about $151.92 each year just by cutting the amount of time that the oven is used from 100 hours to 50 hours each month by using smaller appliances.

3. Ditching the Long Hot Showers

Now we have all heard the age-old debate about showers being more water-conscious than baths. But to what point, as showers are constantly draining and bathtubs hold water?

According to the US Geological Survey, the average shower is between 2 and 5 gallons per minute (7.6 and 18.9 L per minute), and the average bathtub holds 36 gallons (136.3 L) ("Water Questions & Answers, n.d.). So how long do you need to be in the shower before you use more water than the bathtub will hold?

Average shower flow rate = 13.25 L/min

In 5 minutes = (13.25 L / min) x 5 min = 66.25 L

In 10 minutes = (13.25 L / min) x 10 min = 132.5 L

In 15 minutes = (13.25 L / min) x 15 min = 198.75 L

So in 15 minutes, you have consumed more water than you would in a bath. Also, it would not be unreasonable to say that someone could stay in the bathtub for 20 to 30 minutes. Therefore, there's no need to feel guilty after a hard day at work about enjoying a nice, soothing bath rather than taking a long, hot shower. But if you can shower in 5 to 10 minutes, then skip the bath and save some water.

4. Shopping Local

I know a lot of us have heard to shop local for your groceries to reduce your carbon footprint. But what does that exactly mean? As we all know, we are very fortunate to have a wide variety of produce available to us year-round. Of course, many of them, such as bananas around in the depths of Winter, are shipped from warmer areas on the planet, and many of these means of transportation rely on fossil fuels in order to power them. However, did you know that there is a significant difference in the amount of greenhouse gases that each of these means creates, and it might not be what you think? Thanks to a paper by W. Wakeland et al., we have access to the values of kg of CO2e (carbon dioxide equivalent) per ton-km of various food transportation methods. But what is a ton-km (TKM)?

> *Ton-km*
> *= weight of cargo in tons x distance travelled by the cargo*

For our examples and to keep things simple, we are transporting 1 ton of cargo.

So those grapes that you see in the grocery store are often from Chile. For our example, let's say we need to transport 1 ton of grapes from Santiago, Chile, to Toronto, Canada. In this case, the only feasible way to transport these grapes without them spoiling is to load them onto a plane. From Santiago, Chile, to Toronto, Canada, it is an 8582 km flight.

Ton-km = 1 ton of grapes x 8582 km = 8582 TKM
Planes produce 6.8 kg of CO_2e per ton-km, so:

> *kg of CO_2e*
> *= 6.8 kg per TKM x 8582 TKM*
> *= 58,358 kg of CO_2e was emitted into the atmosphere.*

The average car emits around 6 tons of CO2e each year ('Global Warming and Your Car," n.d.). That single ton of grapes shipped by plane from Santiago, Chile, to Toronto, Canada, emitted the equivalent of what over nine cars emit in a year.

Fortunately, not all cargo is transported by plane. For example, rice shipped from Mumbai, India, to Toronto, Canada, could be shipped by plane, but due to its longer shelf life, it can also be transported by cargo ship. From Mumbai, India, to Toronto, Canada, it is 22,689 km on a shipping route provided by Google, which takes 51 days.

Ton-km = 1 ton of rice x 22,689 km = 22,689 TKM

International Cargo Boats produce 0.14 kg of CO_2e per ton-km, so:

> *kg of CO_2e*
> *= 0.14 kg per TKM x 22,689 TKM*
> *= 3,179 kg of CO_2e*

Whereas we were to fly the rice to Toronto, 85,081 kg of CO_2e would have been produced.

Now it's tough in the wintertime to maintain a balanced diet, especially when there is over a foot of snow outside, and when you're sick, sometimes those navel oranges from Florida make you feel better, and that's ok too. But, with this next example, I am hoping to give you some food for thought.

Summertime in Toronto, and you are making fresh salsa for the neighbourhood BBQ, but you need tomatoes. You have three options:

tomatoes from Mexico City that just arrived this morning at your local grocery store; tomatoes from somewhere in Canada (let's say 1000 km away) at your local grocery store; and tomatoes from your local farmer's market that the farmer travelled 500 km to be there. Which one is the better choice for the atmosphere?

First, let's look at Mexico City Tomatoes:

If by Plane,

Ton-km Fly from MX City to Toronto
= 1 ton tomatoes x 3243 km
= 3243 TKM

kg of CO_2e
= 6.8 kg per TKM x 3243 TKM
= 22,052 kg of CO_2e

If By Truck,

Ton-km
= 1 ton tomatoes x 4097 km
= 4097 TKM

kg of CO_2e
= 1.8 kg per TKM x 4097 TKM
= 7,375 kg of CO_2e

If By Train,

Ton-Km by Train from MX City to Toronto
= 1 ton tomatoes x 4520 km
= 4520 TKM

** Trip by train would take around 5 days to complete, so may be impractical as tomatoes may spoil before reaching Toronto*

kg of CO_2e
= 0.18 kg per TKM x 4520 km

$$= 814\ kg$$

So for the Mexico City Tomatoes, we are looking at 7,375 to 22,052 kg of CO_2e being emitted into the atmosphere.

Next, Tomatoes from 1000 km away:
Plane,

Ton-km
= 1 ton tomatoes x 1000 km
= 1000 TKM

kg of CO_2e
= 6.8 kg per TKM x 1000 TKM
= 6,800 kg of CO_2e

Truck,

Ton-km
= 1 ton tomatoes x 1000 km
= 1000 TKM

kg of CO_2e
= 1.8 kg per TKM x 1000 km
= 1,800 kg of CO_2e

Train,

Ton-km
= 1 ton tomatoes x 1000 km
= 1000 TKM

kg of CO_2e
= 0.18 kg per TKM x 1000 km
= 180 kg of CO_2e

*travelling at approx. 70 km/h, it would be about a 14-hour train trip – feasible means of transport

So for the tomatoes from 1000 km away, we are looking at 180 kg to 6,800 kg of CO_2e being emitted into the atmosphere.

Finally, Tomatoes from 500 km away:
Plane,

> *Ton-km*
> *= 1 ton tomatoes x 500 km*
> *= 500 TKM*
>
> *kg of CO_2e*
> *= 6.8 kg per TKM x 500 TKM*
> *= 3,400 kg of CO_2e*

Truck,

> *Ton-km*
> *= 1 ton tomatoes x 500 km*
> *= 500 TKM*
>
> *kg of CO_2e*
> *= 1.8 kg per TKM x 500 TKM*
> *= 900 kg of CO_2e*

Train,

> *Ton-km*
> *= 1 ton tomatoes x 500 km*
> *= 500 TKM*
>
> *kg of CO_2e*
> *= 0.18 kg per TKM x 500 km*
> *= 90 kg of CO_2e*

**travelling at approx. 70 km/h, it would be about a 7-hour train trip – feasible means of transport*

So for the tomatoes from 500 km away, we are looking at 90 km to 3,400 kg of CO_2e being emitted into the atmosphere.

As you can see, taking the more local options of the tomatoes that are from 500 km and even 1000 km away lessens the amount of CO_2 that is emitted into the atmosphere in comparison to those tomatoes that are transported from further away.

See, it wasn't that hard to help out the Earth, now was it? That's all we have for today. Thanks everyone, for tuning in, and remember, Keep Looking for the Science in Things.

Footnote:

At the time of writing this blog post, the Instant Pot was the must-have kitchen gadget. So much so that I am surprised that I didn't include it in the initial post. Better late than never, as they say:

The Instant Pot (pressure cooker) is 1200 W and is used for about 5 hours each month.

> *Average kWh each month*
> *= (1200 W x 5 hrs) / 1000*
> *= 6 kWh/month*
>
> *Average monthly & yearly energy bill for an Instant Pot*
>
>> *Average monthly $*
>> *= ($0.0614 / kWh) x (6 kWh / month)*
>> *= $0.37/month*
>>
>> *Average yearly $*
>> *= ($0.31 / month) x (12 months / year)*
>> *= $4.44/year*

The Stove/Oven still is 5000 W, but now only 10 hours/month.

> *Average kWh each month*
> *= (5000 W x 10 hrs) / 1000*
> *= 50 kWh/month*
>
> *Average monthly & yearly energy bill for Stove/Oven*
>
>> *Average monthly $*
>> *= ($0.0614 / kWh) x (50 kWh / month)*
>> *= $3.07/month*
>>
>> *Average yearly $*

= ($3.07 / month) x (12 months / year)
= $36.84/year

The Instant Pot, which is 1200 W, is so versatile, and it's safe to use it to cook raw meat (unlike the microwave). Let's say we cook with it for 90 hours a month. So in order to convert from oven hours to Instant Pot hours, we must divide the oven hours by 3.

Instant Pot hours per month (plus the 5 hours we already use it each month)
=(90 hours / 3) + 5 hours
=35 hours/month

Average kWh each month
= (1200 W x 35 hrs) / 1000
= 42 kWh/month

Average monthly & yearly energy bill for Instant Pot

Average monthly $
= ($0.0614 / kWh) x (42 kWh / month)
= $2.58/month

Average yearly $
= ($2.58/ month) x (12 months / year)
= $30.95/year

As a result our total for the year comes to 1104 kWh used for the year, costing a total of $67.89. That's a savings of $300.51/year.

The Lowdown on Wind Turbines

(Originally Published May 9, 2018)

Lately, we as a society have been putting a much-needed focus on diverting our efforts into developing and utilizing greener energy. We have been researching different means to do so, such as wind turbines, solar panels, geothermal, and tidal energy, to name a few. So why not take a look at one of these green technologies? Wind turbines. Certain regions have seen an increase in the presence of wind turbines popping up, especially regions that see consistent wind. They offer a means of producing energy that is greener (there are some environmental costs that are currently being debated) than current methods that are being implemented, such as coal-burning energy plants, petroleum energy, and debatably hydroelectric energy. But how much area does it take to house a wind turbine farm?

For the purpose of this inquiry, let's consider converting Canada to 100% wind turbine energy. So how many turbines would we need to power Canada? And how much land would we need in order to house all the necessary wind turbines?

The average Canadian household consumes 106 GJ/household/year, or 1.06×10^{11} J/household/year (as of StatsCan 2007). According to StatsCan's 2016 count, there were 14,072,080 private households in Canada in 2016.

Total Power Needed
= (1.06 x 10^{11} J/household/year) x 14,072,080 households
= 1.49 x 10^{18} J/year

Power in MW
= (1.49 x 10^{18} J/year) / (365 days)(24 hours)(60 minutes)(60 seconds)
= 4.73 x 10^{10} MW

A commonly used wind turbine is the Vestas V90, which has a rotor diameter of 90 m. These turbines have to be a minimum of 5 rotor lengths from each other. The efficiency of that particular model is as follows:

Power curve

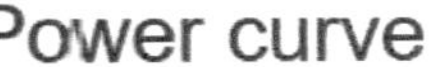

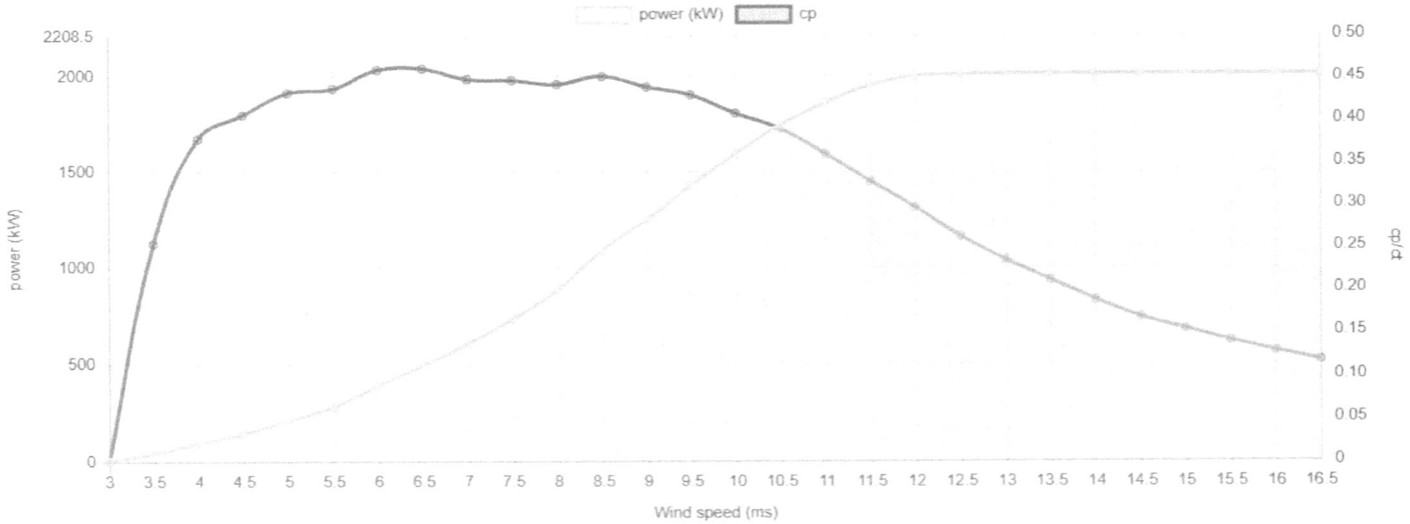
power (kW)
cp
2208.5
2000
1500
1000
500
0
power (kW)
3
3.5
4
4.5
5
5.5
6
6.5
7
7.5
8
8.5
9
9.5
10
10.5
11
11.5
12
12.5
13
13.5
14
14.5
15
15.5
16
16.5
Wind speed (ms)
0.50
0.45
0.40
0.35
0.30
0.25
0.20
0.15
0.10
0.05
0
cp/ct

The average wind speed in Canada is 14 km/h, which is 4 m/s ("Average Annual Wind Speed at Canadian Cities, n.d.); which would result in an efficiency of 0%. Therefore, we would have to be selective in where we place the wind turbines. So for our purposes, let's assume an average wind speed of 12 m/s; the power coefficient (cp), as per the chart above, would be 0.296.

$cp = 0.296$

k_m = mechanical loss of blades and gearbox
        (usually 0 - 0.3%)

k_e = electrical losses of the turbine (usually 1 - 1.5%)

$k_{e,t}$ = electrical losses of transmission to grid
        (usually 3 - 10%)

k_t = percentage of time out of order due to failure or
        maintenance (usually 2 - 3%)

k_w = wake losses due to neighboring turbines & terrain
        topography (usually 3 - 10%)

η = efficiency = $(1 - k_m)(1 - k_e)(1\ k_{e,t})(1 - k_t)(1 - k_w)*cp$
  = $(1 - 0.015)(1 - 0.0125)(1 - 0.065)(1 - 0.025)$
    $*(1 - 0.065)*0.296$
= $(0.985)(0.9875)(0.935)(0.975)(0.935)*0.296$
= 0.245

ρ = air density = $1.2\ kg/m^3$

$A = Area = \pi r^2$

$v = velocity = 12\ m/s$

$$P = (1/2)\rho A v^3$$
$$= \eta(1/2)(1.2\ kg/m^3)\pi(45\ m)^2(12\ m/s)^3$$
$$= (0.245)(1/2)(1.2\ kg/m^3)\pi(45\ m)^2(12\ m/s)^3$$
$$= 1615979.969\ W$$
$$= 1.62\ MW$$

Number of Turbines
$$= 4.73\ x\ 10^{10}\ MW\ /\ 1.62\ MW$$
$$= 29{,}197{,}289{,}710\ turbines$$

Because the wind turbines can be no closer than 5 rotor lengths apart, a straight line of n turbines covers length:

$$L = (n\text{-}1)(5d)$$
$$\qquad where\ n\ is\ the\ number\ of\ turbines\ and\ d\ is\ rotor\ length$$

If the piece of land is square,

$$A = L^2 = 25(n\text{-}1)^2 d^2$$

Therefore, the area of land is:

$$Area = 25(n\text{-}1)^2 d^2$$
$$= 25(29,\ 135{,}802{,}470 - 1)^2(90\ m)^2$$
$$= 1.72\ x\ 10^{26}\ m^2$$
$$= 1.72\ x\ 10^{24}\ km^2$$
$$= 1.72\ \ x\ 10^{22}\ hectares$$

The total Area of Canada is $9{,}984{,}670\ km^2$.

$$1.72\ x\ 10^{24}\ km^2\ /\ 9,984,670\ km^2$$
$$= 172,264,080,800,000,000\ times\ the\ size\ of\ Canada$$

So in order to power Canada entirely on wind turbines, we would need around $1.72\ x\ 10^{22}$ hectares of land, which is 172 quadrillion times the size of Canada.

So what does this mean? We need to further research and improve the efficiency of our wind turbine technology, as we can only achieve a maximum efficiency of around 55%. If that is unattainable, we need to discover a technology that has better efficiency and takes up less space. However, this is not a reason to give up the goal of producing greener energy. In the meantime, we can look at our options to reduce the amount of energy we use.

I hope this has given you some food for thought. In the meantime, Keep Looking For the Science in Things.

Footnote:

The original chart for efficiency from the blog post was no longer available, so a new graph was acquired and the calculations were updated to reflect the new data.

After further researching the Vestas V90 model of wind turbines, it was discovered that the cut-in wind speed is 4 m/s (Appendix J: *V90 3.0Mw Turbine Specifications*). The cut-in wind speed is the speed at which the blades would start turning to generate energy. So although the efficiency would not be 0%, it would be very minimal.

In the article, it is mentioned that further research needs to be done in order to improve efficiency. However, at the time, Betz's Law was not taken into consideration. Betz's Law dictates how much power can be collected from the wind, independent of how the wind turbine is designed. According to Betz's Law, the most efficient that a wind turbine could ever be is 59.3%. This would allow it to generate 3.91 MW, which means Canada would require 12,071,611,250 turbines to meet the power demand, which would require 2.95×10^{21} hectares, which is 295,452,929,300,000 times bigger than Canada.

Kilauea: The Bubbling Giant

(Originally Published May 28, 2018)

So if you have been tuning into the news lately, you probably have heard that for the past month and a bit, the Kilauea Volcano has been erupting, or maybe you saw the time-lapse video of the lava "eating" a car, which went viral at the beginning of this month. Well, today we are going to talk about the Kilauea Volcano and volcanoes in general, in what we are calling Volcanology 101.

So the study of volcanoes, lava, magma, and other related geological occurrences is called volcanology, and the scientists that study volcanology are volcanologists. Not Vulcans, bummer. There are around 500 active volcanoes on Earth, with about 50 to 70 eruptions occurring every year (Pfeiffer, n.d.). Volcanoes are a result of how the Earth is constructed. So let's go back to elementary school science for just a second.

Geology 101

The outer layer of our planet, on which we live, is called the crust. The crust is anywhere from 5 km to 60 km thick at various points around the World (Robertson, 2011). Now, the crust is not a solid shell; it is made of sections called tectonic plates. There are seven major tectonic plates and 152 smaller plates that make up the Earth's crust (Harrison, 2016). These plates float on the magma that makes up the Earth's mantle. The Earth's mantle has a current that is always moving, and as a result, the tectonic plates move as well. This movement is most noticeable at the plates' edges, in areas that we call fault lines. Whether it is past each other (like off the coast of California), one over top of the other (like in the Himalayas), or both away from each other (like in the middle of the Atlantic Ocean), because of this movement, magma often escapes from the mantle and bubbles up to the surface. In many of these occurrences, volcanoes form. In fact, the majority of volcanoes occur on active fault lines (Lynch, n.d.). But there is one area on the Earth where tectonic plate movement is much more active and, in some cases, much more volatile. That area is referred to as the Ring of Fire and it circles the outside parameter of the Pacific Ocean, and holds 75% (which is more than 400 volcanoes) of Earth's volcanoes ("Ring of Fire," n.d.).

Now, the volcanoes in Hawaii are different from those that are part of the Ring of Fire. Located in the middle of the Pacific Ocean Tectonic Plate, there are no fault lines for magma to bubble up. In the case of Hawaii, magma broke through the Earth's crust in order to form the volcanic chain of islands ("How did the Hawaiian Islands form?" n.d.). When this occurs, it is referred to as a hot spot.

Types of Volcanoes

There are seven types of volcanoes and one major volcanic formation worth discussing today. They are fissure vents, shield volcanoes, stratovolcanoes, supervolcanos, submarine volcanoes, subglacial volcanoes, mud volcanoes, and lava domes.

Fissure Vent

A fissure vent is a linear vent on the ground in a volcanic area where lava erupts from ("Types of Volcanoes," 2013). Not usually explosive, fissure vents vary in size from a few meters to several kilometres in length. However, they can be difficult to recognize unless you are right above them, as they are generally flat, with no dome or cone shape. Unlike other volcano types, it has no central caldera, or depression, making it quite easily covered up during an eruption.

Shield Volcano

A shield volcano, much like its name implies, is shaped like a shield. It is formed due to several lava flows, which give it its gradual shield shape ('Types of Volcanoes," 2013). Because of how they form, they are very wide and broad, with bases several kilometres in diameter, a steeper middle slope, and a flatter summit (Bagley, 2018). Their eruptions are not generally explosive, with magma usually oozing out of them slowly (Bagley, 2018). The famous volcano Mauna Loa is a great example of a shield volcano, as well as being the World's largest active volcano ("Mauna Loa," n.d.).

Stratovolcano

Stratovolcanoes are your stereotypical volcano, very tall/steep (many exceeding 2500 meters) and have the characteristic cone shape ("Types of Volcanoes," 2013). They are built up from alternating lava flows, ash, and blocks of unmelted stone that form around a system of vents that lead to a magma reservoir under the Earth's surface (Bagley, 2018). Eruptions tend to be rather violent due to a buildup of pressure in the magma chambers (Bagley, 2018). During an eruption, lava shooting out tends to be very violent and fast, but because of this action, the lava cools faster than other volcano types, so it doesn't travel as far. Because of all the built-up pressure, it is not uncommon for the sides of the cone and the summit crater to blow out (Bagley, 2018). Examples of famous stratovolcanoes that you may be familiar with include Mount Fuji, Mount Vesuvius, Stromboli, and Mount St. Helens.

Supervolcano

You may have heard about supervolcanoes ("Types of Volcanoes," 2013). These types of volcanoes are known to cause the most damage ("Types of Volcanoes," 2013). Some scientists believe that a supervolcano eruption could end life on Earth as we know it ("Types of Volcanoes," 2013). The last supervolcano eruption was the Yellowstone supervolcano, which many scientists believe sparked the last Ice Age, which biologically,

environmentally, and geologically altered our planet, wiping out
many species of animals ("Types of Volcanoes," 2013). There
are six famous supervolcanoes that are currently being
monitored and studied ("Types of Volcanoes," 2013). They are
Taupo Caldera in New Zealand, Aira Caldera in Japan, Lake
Toba in Indonesia, Yellowstone Caldera, Valles Caldera, and
Long Valley Caldera, all of which are in the US.

Submarine Volcano

Submarine volcanoes are fissures or vents found underwater that
erupt magma ("Types of Volcanoes," 2013). Approximately
75% of the magma erupted every year comes from submarine
volcanoes ("Types of Volcanoes," 2013). They are mostly located
in areas of high tectonic movement (the edges of the tectonic
plates) and are found primarily at great depths, although there
are a few that have been discovered closer to the surface ("Types
of Volcanoes," 2013). Because of their location, the magma
during an eruption is cooled rapidly due to immediate contact
with cold, deep ocean water. However, this poses a risk of
flooding and, in extreme cases, tsunamis during an eruption
because of the water displacement created by the addition of
volcanic material and earthquake activity prior to the eruption.

Subglacial Volcano

Subglacial volcanoes are found under glaciers and ice sheets
("Types of Volcanoes," 2013). They are capable of quickly
forming lakes due to the extreme heat produced during
an eruption ("Types of Volcanoes," 2013). Mostly found in
Antarctica and Iceland, some have been found in the Yukon and
British Columbia, Canada ("Types of Volcanoes," 2013). Similar
to submarine volcanoes in that magma cools rapidly during an
eruption, only this time due to contact with the ice and water
formation. Areas near subglacial volcanoes are at risk of

flooding during eruptions due to the excess water produced and the water displacement caused by the additional volcanic material ("Types of Volcanoes," 2013).

Mud Volcano

Mud volcanoes, also called mud domes, form from a buildup in pressure of various gases and liquids under the surface ("Types of Volcanoes," 2013). When the hot water and mud mix with the surface soils, they push up the surface, creating bubbles or domes in the landscape ("Types of Volcanoes," 2013). Commonly found in subduction zones, or where one tectonic plate goes under another ("Types of Volcanoes," 2013). There are currently around 1100 of these being watched and studied ("Types of Volcanoes," 2013). Mud volcanoes are known to generally indicate the presence of a possible petroleum deposit or a nearby volcano ("Types of Volcanoes," 2013). Those near volcanoes emit helium, whereas those that are not emit methane. The current estimate of the number of mud volcanoes in existence are around 10,000.

Lava Domes

Lava domes are volcanic structures rather than a type of volcano ("Types of Volcanoes," 2013). They are circular formations due to highly liquid lava building up because it is too viscous to move ("Types of Volcanoes," 2013). They are also known to form when there is not enough pressure for an explosion to occur, so the lava just oozes out of the vent. Then a bubble or plug of cooling rock forms over the fissure or vent, creating a lava dome (Bagley, 2018). They are commonly seen in the craters of stratovolcanoes, an example being Mount St. Helens, which has several well-formed lava domes (Bagley, 2018).

Dangers of Volcanoes

When discussing the dangers of volcanoes, many come to mind, such as lava, earthquakes, tsunamis, floods, and mudslides. Although lava is highly destructive to everything it touches, it generally does not move that rapidly, so many are able to get out of its way ("Lava flows," n.d.). But one danger that many underestimate, as the people of Pompeii certainly did, is the pyroclastic flow ("Lava flows," n.d.). The pyroclastic flow is the cloud of poisonous gas, ash, and debris that is launched into the air during the eruption, particularly in more violent eruptions. The ash cloud can travel up to 160 km/h, bulldozing anything and everything that gets in its path as it races down the volcano's slope. Rocks that are launched, ranging from small pebbles to bus-sized boulders, have been measured to travel at speeds of 80 km/h (Volcano Hazards, n.d.). And if you can avoid the poisonous gas and flying boulders, temperatures inside the pyroclastic flow range from 200 to 700 degrees Celsius (Volcano Hazards, n.d.). Ouch! So, if you feel the ground rumble, maybe check to see if you need to get out of the area.

Kilauea

Now, what about Kilauea? Well, Kilauea is classified as a shield volcano. In fact, it is the youngest (on land) shield volcano in all of Hawaii (*Volcano Discovery*, n.d.). And it is the World's most active volcano (*Volcano Discovery*, n.d.). Located at the southern end of the Big Island of Hawaii, it is located relatively near Mauna Loa, despite being its own separate volcano (not a satellite of Mauna Loa, as previously believed) (*Basic Planet*, n.d.). Its main central depression measures 3 km by 5 km and is believed to have formed in several stages about 1500 years ago (*Volcano Discovery*, n.d.). Measuring 6 km by 6 km at its outermost faults, Kilauea goes down 165 meters deep and 1277 meters high, with a magma system that extends over 60 km deep into the Earth's mantle (*Volcano Discovery*, n.d. | *Basic Planet*, n.d.). About 90% of the shield's surface is formed from lava flows that are less than 1100 years old, with 70% being within the past 600 years (*Volcano Discovery*, n.d.).

Scientists speculate that Kilauea first began erupting between 300,000 and 600,000 years ago (*Basic Planet*, n.d.). Because it is a shield volcano, Kilauea's eruptions are not particularly violent. In fact, Kilauea's name comes from the Hawaiian word meaning "much spreading" or "spewing", a testament to its frequent lava flows (*Basic Planet*, n.d.). Currently, it is undergoing one of the longest-lived eruptions known on Earth, which started on January 3, 1983, and is continuing today (*Volcano Discovery*, n.d. | *Basic Planet*, n.d.). Although it started from the central vent, the lava exit points have shifted many times over the years, with the most recent activity being this past month (*Basic Planet*, n.d.). Its usual path flows through two rift zones to the East and Southwest of the volcano, extending all the way to the ocean (*Volcano Discovery*, n.d.). Lava flows vary from day to day, from 300,000 to 600,000 cubic meters (that is enough to fill 120 to 240 Olympic-sized swimming pools) to days of no eruptive activity at all (*Basic Planet*, n.d.). Since this current eruption started, more than 100 square kilometres have been covered, and around 200 homes have been destroyed (*Volcano Discovery*, n.d.). However, it is not all bad news because, thanks to Kilauea, the coastline of the Big Island of Hawaii is expanding, allowing more visitors to enjoy the beautiful, ever-expanding beaches of Hawaii.

So, this month's news is not necessarily a new eruption by Kilauea, but rather the most recent change in activities from the volcano, with the last notable amount of activity being 7 years ago. With this recent shift towards human settlement, Kilauea has reminded the world who really is in charge. The Earth. For further updates and information on the Kilauea Volcano and other Hawaiian volcanoes (including Mauna Loa, which is being closely watched), head over to the US Geological Survey's site. And that's all for today's post. In the meantime, Keep Looking for the Science in Things.

Footnote:

Kilauea's eruption that was observed in 2018 ended on September 5, 2018 (Smithsonian Institution, n.d.). Two more confirmed eruption events have occurred since then. The first one was from December 20, 2020, to May 23, 2021, during which there was a lava lake and fountains recorded (Smithsonian Institute, n.d.). The second event began at the Halema'uma'u Crater on September 29, 2021, and is considered ongoing (Smithsonian Institution, n.d.).

Sadly, on December 9, 2019, the Whakaari volcano near New Zealand erupted, claiming the lives of 22 people, mostly tourists (Britannica, 2023). The volcano had been exhibiting heightened volcanic activity a few weeks prior to the incident, but there were no indications that an eruption was to be expected so soon (Encyclopedia Britannica, 2023). Scientists believe that water came into contact with the hot magma at the volcano's core, causing steam to form and pressure to build until the volcano erupted (Encyclopedia Britannica, 2023). It is events like this that remind us to never underestimate volcanoes and their capabilities.

Hurricanes, Cyclones, Typhoons... Oh My!

(Originally Published May 30, 2018)

Well, tropical storm season has once again begun in the Caribbean and Southeastern United States, with tropical storm Alberto making its way towards the state of Florida. For the many people that live in the Gulf of Mexico and up the Eastern Seaboard of North America, discussions of hurricanes and tropical storms are part of life. But for the rest of us, the only time that we see mention of a hurricane or a tropical storm is in the news. So what is a hurricane? How is it different from a tropical storm? Is it the same as a cyclone or a typhoon? What is their season? These are among the questions that we will be discussing in today's post.

What is a Hurricane?

Hurricanes are rotating systems of weather made up of clouds and thunderstorms that produce strong winds and heavy rains. They form in warm tropical and subtropical waters when there is a significant difference in temperature and air pressure between the ocean water and the clouds above (*National Ocean Service*, n.d.). As per the water cycle, the clouds and atmosphere pull up moisture from the ocean, but because of the strong difference in temperatures and pressures, a fast-moving column of air also forms. As the warm, moist air rises, it leaves behind a low-pressure area where more warm, moist air rushes in to fill it. This air is then pulled up into the building storm system as well, with surrounding air swirling in to fill the void (swirling counterclockwise if it was formed north of the Equator, and clockwise if it was formed south of the Equator) (*NASA Space Place, n.d.*). As the storm builds, it increases in rotation speed and begins to form the hurricane's characteristic eye (with the cold, high-pressure air dropping down in the centre of the eye and the warm, low-pressure air rising and swirling in the "walls" of the hurricane). Hurricanes get their energy from the warm, moist air, so their movement follows warm water and air currents (Brain et al., 2000). This is also why hurricanes quickly lose momentum after they hit land or venture into cooler, northern waters.

Measuring a Hurricane

Hurricanes are measured on the Saffir-Simpson Hurricane Wind Scale (SSHWS), which is based on the Beaufort Wind Scale (Belles, 2017). The SSHWS is broken down into five categories, which are the categories of a hurricane. You may have heard of a category 4 hurricane on the news. These categories are based on the hurricane's sustained wind speed ("Saffir-Simpson Wind Scale," n.d.). And they are as follows:

Category 1	119 – 153 km/h
Category 2	154 – 177 km/h
Category 3	178 – 208 km/h
Category 4	209 – 251 km/h
Category 5	252 km/h and higher

In addition to their wind speeds, each category is also characterized by the amount of damage that occurs.

Hurricane vs. Tropical Storm

During this time of year, it seems like the words hurricane and tropical storm are tossed around a lot. But what is the difference? It all comes down to wind speed (Belles, 2017). Hurricanes are measured on the Saffir-Simpson Hurricane Wind Scale, which starts at 119 km/h. So if a tropical storm is travelling between 63 km/h and 117 km/h, it is classified as a tropical storm, not a hurricane. And below 63 km/h, it is referred to as a tropical depression ("Tropical Cyclone Climatology," n.d.).

Hurricanes, Cyclones, and Typhoons… Oh My!

But what about the other names: cyclones and typhoons? Well, a tropical cyclone is a general term that refers to the weather phenomenon also known as hurricanes and typhoons ("What is the difference between a hurricane and a typhoon?" n.d.). In other words, all hurricanes are cyclones, but not all cyclones are hurricanes. The difference between hurricanes and typhoons is where they form and occur. Hurricanes, as many of us know, form in the North Atlantic as well as the central North Pacific and Eastern North Pacific; whereas, typhoons occur in the Northwest Pacific. They both form the same way and have the same minimum wind speed of 119 km/h.

Hurricane Season

Hurricane season in the Atlantic generally occurs between June 1 and November 30; however, hurricanes can occur outside of those dates. As for the Eastern Pacific, hurricane season is generally from May 15 to November 30 ("Tropical Cyclone Climatology," n.d.). Typhoons, on the other hand, have no specified season, as typhoons can occur at any time (the majority occurring between mid-February and November, with the peak from July to November) ("When is hurricane season?" n.d.).

Cyclone Hazards

As we approach tropical storm and cyclone peak season and as people venture out on Summer vacations, there are some hazards that may not be apparent or widely known. Even those of us who grew up outside of hurricane/typhoon area are aware of the high winds, extensive rainstorms, and flooding, but there are some other risks as well:

Storm surges are abnormal rises in the water due to the storm's winds pushing the water towards land ('Storm Surge Overview," n.d.). They are capable of travelling great distances inland, and they are the single greatest cause of loss of life and destruction of property.

Storm Tides, on the other hand, are a combination of the astronomical tide (the tide that is controlled by the Moon) in addition to the rising waters of the storm surge ('Storm Surge Overview," n.d.). They can be as damaging as storm surges.

Tornadoes (this is one that I am finding out about for the first time as well) can accompany hurricanes once they make landfall ("Hurricane Safety Tips and Resources," n.d.). Although not in the immediate vicinity of the eye of the hurricane, tornadoes have been known to occur within the hurricane's rain bands, wreaking additional havoc.

Tropical Storm Safety

So what can you do to stay safe? Well, there are many tips offered by government agencies, and we have gathered them here to keep you safe at home and on vacation:

- Check the area you live in or are visiting to see if it is part of the hurricane evacuation zone.

- Be sure to put together an emergency kit, which can include:

- 12 litres of water per person per day for a minimum of 3 days (so the average 4-member household will have a minimum of 144 litres)
- Water purifiers
- Minimum 3-day supply of non-perishable (canned/dry goods) food
- Battery-powered or hand crank radio
- Flashlight
- First aid kit
- Extra batteries
- Whistle to signal for help (flares and air horns are nice, but they have limited use)
- Dust masks
- Plastic sheeting and duct tape (for temporary shelter)
- Moist towelettes and garbage bags for personal sanitation (which we rarely think about with the convenience of indoor plumbing)
- Crescent wrench, multitool, or pliers
- Manual can opener
- Local maps (non-electronic)
- Cell phone with charge cords and backup battery/portable charger
- Prescription medications
- Non-prescription meds (e.g. Advil, Tylenol, Pepto-Bismol, etc.)
- Glasses and contact lens solution
- Infant necessities (diapers, formula, wipes, etc.), if needed
- Pet food and extra water, if needed
- Cash and traveler's cheques
- Important documents should be saved to an external hard drive or fire resistant and waterproof container
- Sleeping bag or warm blanket for each person
- Climate-appropriate change of clothes (per person)
- Sturdy shoes (per person)

- Fire extinguisher
    - Matches in a waterproof container
    - Feminine supplies and personal hygiene items
    - Basic mess kit and camping dinnerware

- Have an emergency plan and make sure everyone knows it. Practice it a few times.

- Double-check your emergency equipment before there is a storm. That includes your flashlights, generators, storm shutters, and water pumps. It's better to find out if they need repairs before the storm hits.

- Stay tuned to the weather reports.

So this Summer and during cyclone season, stay safe and be sure to have fun. As long as you stay mindful of the weather, you should have a fun, sun-filled Summer and vacation. And don't forget the sunscreen. That's all for today's post, from Schrodinger's Cat, "Keep Looking for the Science in Things."

Schrodinger's Cat's Quick Guide to Climate Change

(Originally Published June 4, 2018)

Discussion about global warming and climate change is one of the hot-button topics that's been in the news over the past decade. With people on either side of the discussion of whether or not it is occurring and who or what is, in fact, responsible for it. Well, there is more than enough evidence to prove that it is not a hoax created by the Chinese, as some would believe. But what is the big deal with global warming and climate change anyway?

Greenhouse Effect

Ever stood in a greenhouse and wondered how it could be so warm, even though it is made of glass or thin plastic? Solar radiation (UV rays) from the Sun radiates down and hits the greenhouse ("Greenhouse effect, n.d.). Some of the radiation is deflected off the greenhouse, but some also enters the greenhouse. That energy is absorbed by everything in the greenhouse: the dirt, the plants, the tables, etc. This absorbed energy is radiated back out as infrared energy. However, this energy is a different wavelength (shape) from the solar radiation, so some of this energy is not able to leave the greenhouse and is deflected back into the greenhouse, which raises (maintains) the temperature in the greenhouse. As more solar radiation shines into the greenhouse, more infrared energy is created in the greenhouse, and the temperature is raised (maintained). Now, what if I told you Earth was just a giant greenhouse? Only instead of glass or plastic, the Earth's atmosphere holds in some of the infrared energy, which maintains the temperature necessary for life.

Global Warming vs Climate Change

So what is climate change anyway? And what about global warming? Lately, we have seen both of these terms used interchangeably, but there is actually a difference. Climate change is the shift in overall weather conditions over an extended period of time ("Frequently asked questions about climate change," 2015). These changes include temperature, precipitation, winds, and weather phenomena and anomalies such as extreme drought, increased number and severity of wildfires, uncharacteristic flooding, increased number and severity of hurricanes, tornadoes, etc. It also varies depending on the location. So if the Sahara desert experienced snowfall on a more regular basis, say annually, that would be considered climate change for that region. Global warming, on the other hand, is an indicator of climate change ("Frequently asked questions about climate change," 2015). As the climate varies depending on the region, global warming will vary as well. However, it is measured as an increase in the global average surface temperature over an extended period of time.

If we are experiencing Global Warming, then why was last Winter so cold?

Well, how cold was last year, really? Was it colder than last year or the year before? How about 10 years ago, or even 50 or 100 years ago? A lot of the time, what we think is cold really isn't if we look at the long-term trends in temperature. Scientists look at the long term because it takes out the anomalies that would otherwise impact their data and give false information. But last Winter, we set a record for the coldest year in thirty years! Well, that's an example of a weather anomaly, but what if I told you that technically we are supposed to be in an ice age? Does that change your opinion on global warming? That's right, we are supposed to be in the middle of an ice age right now. And this is how I know. Many experts claim that the Earth goes through climate change cycles that shift from snowball Earths (where the planet is frozen from pole to pole) to very tropical climates (similar to when dinosaurs roam the

Earth) (Watkins, n.d.). And technically, right now, we are in an ice age, that's why we have polar ice caps. It's a little scary to think that technically, we are supposed to be much cooler than we actually are (Geggel, 2017).

What's the big deal if it gets warmer?

If it gets warm, then we don't have to shovel snow in Winter. We could wear shorts and sandals all the time. So what's the big deal if the planet gets a little bit warmer. Well, as much as I hate being cold, there is a reason why Earth's average temperature is around 15 degrees Celsius and not in the high twenties ("Global Warming," 2010). First of all, the rise in average temperature is jeopardizing our polar ice and other landlocked ice. When all that ice melts, the water has to go somewhere, and that eventual somewhere is our oceans. Can't afford oceanfront property right now? Just wait for the ice caps to melt, and your backyard just might become the ocean, or be at the bottom of it. Scientists speculate that the ocean could rise as much as 95 cm (almost a meter) (Brain, 2000). And rising ocean levels are just the start. In addition to rising ocean levels, the ocean temperature and that's not a good thing. Marine life has evolved over time within a small range of temperature variations. By raising temperatures out of these ranges (which has been happening more and more frequently for longer and longer timeframes), it places marine life under stress, makes ocean waters more acidic and less oxygenated. This stress can be seen in coral bleaching, marine species seen outside of their traditional range, and mass species die-off. And the ocean isn't the only place that will be affected by the increase in temperature. On land, we will see the amount of land that we can still farm continue to shrink. Agricultural production will decrease, and we will experience a food shortage like we have never seen before (Cline, 2008).

What can be done to fix this?

So what can we do to fix this problem? Well, some scientists claim that we have passed the point of no return, but that doesn't mean we should give up. There is plenty we can do in order to maintain the Earth at the point we are at now. In addition to the suggestion in my post, 4 Ways to Make the Planet Greener, While Saving You Some Green, there are plenty of other ways that we can help lessen our impact on climate, and our planet as a whole, including:

- Sign up for greener energy options. See where your electricity is actually generated - is it coal-based or hydroelectric? Maybe it's solar or wind-generated. Being informed is half the battle.

- Recycle as much as you possibly can. You physically drag less to the curb each week in your garbage cans and help to reduce the amount of energy and resources spent to extract and process the raw materials used to make that cardboard box or plastic bag.

- Purchase sustainable natural products when you can. Rather than buying polyester clothing and linens, check out sustainable bamboo products. They are softer and breathe better, in addition to being much easier to process and make than your plastic clothes (yes, polyester is plastic).

- Buy local, organic products. In doing so, you cut down on the emissions created in the transportation of these goods as well as the pesticides and herbicides used.

- Invest in reusable products, such as reusable grocery bags, reusable drinking straws, and reusable hot and cold beverage cups. By doing so, you reduce the amount of plastic that goes to landfills, as well as the energy cost of extracting and processing the raw material to make new disposable products.

So we may be in a serious predicament, but there is still plenty that we can do to keep it from getting any worse. Climate change is certainly no joke or hoax. We only have one Earth, so we should do everything we can to take care of it. And that starts with education, especially for people in influential positions like presidents and prime ministers. That's all for today's post, and remember, Keep Looking for the Science in Things.

<u>Footnote:</u>

Unfortunately, since 2018, we haven't seen much improvement. Global temperatures continue to rise. 2020 tied with 2016 for the hottest year on record since record keeping began in 1880, with an annual average global temperature of 1.02 °C ("Vital Signs: Global Temperature, 2023). Breaking it down by month, according to the European Centre for Medium-Range Weather Forecasts (ECMWF) and the Copernicus Project, the hottest years on record are:

- January 2020
- February 2016
- March 2016
- April 2016
- May 2020
- June 2023
- July 2019
- August 2016
- September 2020
- October 2019

- November 2020
- December 2019

To see that all the hottest months on record have happened in the last seven years is concerning, to say the least. And the rise in surface temperature is the only thing that we mentioned in 2018 that continues to be concerning. 2022 marked thirty years since TOPEX/Poseidon was launched into space in an effort by scientists to better monitor ocean surface height. Due to the collected data, scientists have not only been able to confirm that our oceans are rising, but the rate at which they are rising is also increasing. In the early 1990s, the oceans were rising at a rate of 2.5 mm/year, which further increased to 3.9 mm/year in the 2010s (Carlowicz, n.d.). In the thirty years since TOPEX/Poseidon was launched, the world's oceans have risen 10.1 cm (Carlowicz, n.d.). Bear in mind that over the past 140 years, the oceans have risen a total of 21 to 24 cm, meaning almost half of that has been in the past thirty years. If that wasn't concern enough, scientists have also stated that for every 2.5 cm that the oceans rise, approximately 2.5 meters of beachfront is lost along the average coastline (Carlowicz, n.d.).

However, we have some light at the end of the tunnel. As of 2019, 170 nations have pledged to ban single-use plastics by 2030 (*World Economic Forum*, 2020). Although this may not seem like a big enough action, all things considered, this will greatly reduce the amount of energy used to extract and process the resources needed in order to make single-use plastics, as well as the pollution created from the disposal of single-use plastics.

6 Ocean Habitats You've Never Heard Of

(Originally Published June 6, 2018)

World Ocean Day is this Friday, June 8, 2018, and there is no doubt that we need to take better care of our oceans, especially considering scientists speculate the amount of plastic in the oceans will outweigh the entire fish biomass by 2050 (Chan, 2016). As we speak, there are THREE trash islands that have formed out in the middle of our oceans, and they are growing every day (Andrews, n.d.). More than 100,000 sea turtles and birds die from ingestion and entanglement from our mismanaged garbage every year (Andrews, n.d.). It's disgusting. To make matters worse, we are still discovering marine ecosystems in these areas that we did not even know existed, and we have no way of knowing how our actions even affect them. We have all heard of kelp forests, coral reefs, and tidal pools, but today, we will be looking at six different, fascinating marine ecosystems (and frankly, some of my favourites from my undergrad) that you may not have even heard of yet.

1. Antarctic Ice Shelves

This is a marine ecosystem that is still being extensively studied today, as our opportunities to view the ocean floor below the towering Antarctic ice shelves are very few and far between. Unlike other ecosystems where the keystone species (the species that holds the entire ecosystem in balance) is an apex predator, such as wolves in Yellowstone National Park and sea otters in kelp forests, the Antarctic ice shelves' key species is the Antarctic krill ("Krill Hotspots: About, n.d.). These tiny crustaceans feed everything from fish to behemoth whales and travel around in dense clusters searching for their next meal, which happens to be phytoplankton (photosynthesizing plankton) and green algae that cling to the underside of the ice shelves and sheets ("Animals of the Ice," n.d.). As our polar ice caps continue to melt, scientists are in a race against the clock to study this threatened ecosystem before it disappears with the ice.

2. Fjords

Fjords, commonly found in the Northern Hemisphere (in areas such as Northern Europe, Canada, and Greenland), are inlets that were formed from glacier activities, and as a result, they get their characteristic appearance of being long (reaching hundreds of kilometres inland), narrow, and often very deep ("Fjord," n.d.). Unlike other inlets that have sloping shores, fjords commonly have steep rock sides that plunge deep below the water's surface. Some of the deepest fjords reach depths of around 2000 meters ("Fjord," n.d.). Because of these unique characteristics, it is unlike any other marine ecosystem in the World. It is commonly populated by deep-sea organisms, deep-water coral reefs (which are home to very old, slow-growing corals; older than tropical reefs like the Great Barrier Reef), orcas, whales, Greenland sharks, seals, and salmon, all of which are capable of venturing far inland in the fjords without leaving the deep, cold waters ("Fjord," n.d.). Fjords are commonly classified as estuaries as they do not "refresh" their waters as frequently as other inlets, which makes them particularly vulnerable to the damage of pollution.

3. Hydrothermal Vents

Hydrothermal vents, which we briefly touched on in the Kilauea post, occur at the bottom of the ocean along the edges of the tectonic plates as well as make up undersea mountain ranges, seamounts, and mid-ocean ridges ('Deep Hydrothermal Vent," n.d.). They are also referred to as white or black smokers because of the high concentrations of minerals (mainly sulfur compounds) in the hot water that is leaving the vents. Because of how deep these vents are, none of the creatures that call hydrothermal vents home rely on photosynthesis for nutrients. Rather, there are bacteria that convert sulfur into food and energy, which in turn are eaten by larger organisms ("Deep Hydrothermal Vent," n.d.). These organisms include giant tube worms, deep-sea mussels, and yeti crabs. Some scientists believe that life on Earth started at these hydrothermal vents ("Deep Hydrothermal Vent," n.d.).

4. Seamounts

Seamounts are exactly what they sound like undersea mountains. They are formed through volcanic activity. They make up about 28.8 million square kilometres, which is larger than any of the land-based ecosystems ("What is a seamount?" n.d.). Because of their structure and shape, they allow for an upwelling of nutrients and food, which supports a wide array of marine species. They are also home to unique corals, some of which are only found on that particular seamount, and no other ("Seamounts," n.d.). The current number of different types of species that are supported by this ecosystem is as high as 1,300 ("Seamounts," n.d.). Unfortunately, because of this rich diversity, it is also a prime target for commercial fishing and trawling, which scoop up everything in their path, including the coral reefs.

5. Cold Seeps

Cold seeps are a unique ecosystem. Much like hydrothermal vents, they are vents; however, instead of hot water, cold seeps release cold, methane-rich water that sits like a lake at the bottom of the ocean ("Deep Hydrothermal Vent," n.d.). Again, like at hydrothermal vents, bacteria convert methane into food and energy, and the bacteria, in turn, feed the other species that live around the cold seep. These species include mussels, shrimps, squat lobsters, tube worms, cutthroat eels, and polychaete worms ("Cold Seeps," 2002). Other than the species that live around the edge of the cold seep, not much else can survive here, especially in the cold seep itself, due to its toxicity (*BBC Earth*, 2007).

6. Glass Sponge Reefs

Thought to have gone extinct not long after the dinosaurs, glass sponge reefs were shockingly discovered off the BC coast (in Canada) in 1987, in the Hecate Strait ("The Discovery," n.d.). To date, around 26 of these prehistoric reefs have been discovered, all along the coast of BC and Alaska, with the Hecate Strait reef estimated to be around 9,000 years old ("The Discovery," n.d.). Each of these reefs are unique in species, shape, and size from the other, making them truly one-of-a-kind ("The Discovery," n.d.). They are made entirely of silica, which is where they get their name "glass" from, making them quite delicate and sensitive. As a result, they are vulnerable to destruction by heavy fishing gear crushing them as well as being choked out by pollution ("The Discovery," n.d.).

Many of the ecosystems we discussed today have limited information about them because they have only been on our radar for 30 or 40 years. Scientists are racing against the clock to get as much information as they can before these ecosystems are altered by human contact or wiped off the planet entirely. And to think, we have only explored 5% of our oceans ("How much of the ocean have we explored?" n.d.). What uniquely beautiful ecosystems are slipping away from the 95% without our knowledge? That we will never know. So this World Oceans Day, do your part and recycle, buy reusable, or even learn a little bit more about the deep blue. That's all from Schrodinger's Cat today; just remember: Keep Looking for the Science in Things.

<u>Footnote:</u>

Since the original blog post was written in 2018, a few things have changed:

- Scientists have now identified five major garbage patches, each located at one of the oceans' main gyres, three of which are located in the North Pacific Garbage Patch (*NOAA Ocean Podcast,* n.d.).
- A new habitat was discovered in 2020 teeming with tiny, shrimp-like crustaceans, 500 meters under the Larsen Ice Shelf, on the eastern coast of the Antarctic Peninsula (*NIWA,* 2022).

- In March 2023, for the first time in more than 40 years, scientists at the Schmidt Ocean Institute discovered three new active hydrothermal vent fields, spanning almost 700 km along the Mid-Atlantic Ridge (*Schmidt Ocean Institute*, 2023).
- As of January 2022, the Department of Fisheries and Oceans Canada has banned bottom fishing in Howe Sound near all prehistoric and living glass-sponge reefs, according to the Canadian Parks and Wilderness Society.

Shining A Light on Solar Energy

(Originally Published June 21, 2018)

Since our post discussing wind turbines was so popular, I've decided to write the next installment in the green energy series we have started. Another up-and-comer in the world of green energy, solar energy has greatly diversified over the past few years in the types of panels that are available. Originally, solar panels were only available in a size that could be summed up in two words: massive and industrial. Today, solar panels come in all shapes and sizes and are even being incorporated into roofing tiles. But how many panels would you need, and how much space would they take up? For an individual home, it can be quite easily contained on the house's rooftop. But what if we wanted to power an entire metropolitan area with solar? Well, I've crunched the numbers, and this is what I got...

Before I start showing you the math and how it all adds up, there are a few things that I need to make clear. First of all, the amount of sunlight (or solar energy) that hits the Earth is 1370 watts per square meter. That is on a 100% clear day, with not a cloud in the sky. Of course, not every day brings bright blue, beautiful skies, and that's ok because solar panels do work with partial sunshine. It's just not as efficient. In my calculations, I do take this cloud cover efficiency into account. Secondly, in my calculations, I am assuming that the solar panels are 100% efficient. What I mean by that is that every photon of solar energy that hits the panels will be converted into usable electrical energy. This is not realistic, and efficiency varies from panel to panel, but it will give us an idea of the minimum amount of land that we will need in order to power up these metropolitans. Finally, I have chosen nine cities across Canada to be our "guinea pig" metropolitan areas. One thing to bear in mind with these locations and locations similar to these is that, despite the cloud cover, these areas also experience winter weather. So my calculations assume that they stay free of snow and ice and will be kept in perfect working order year-round. This is not always possible in cities such as Edmonton, Saskatoon, and "Winterpeg." So the ability to keep the panels clear and operational is something to consider before switching a city to strictly solar. Now that we have that out of the way, on to the math.

The Canadian metropolitan areas that we will be looking at are Vancouver, Edmonton, Saskatoon, Winnipeg, Toronto, Montreal, St. John's (Newfoundland), Halifax, and Moncton. Below I have made a chart of each city, their area (in square meters), their populations, the average percentage of sunlight they get in a day (cloud cover efficiency), and the average number of hours of sunlight they get each year (*Current Results,* n.d.).

City	Area (in sq. m)	Population (as of 2014)	Cloud Cover Efficiency	Hours of Sunlight
Vancouver	2,882,680	647,540	41%	1938
Edmonton	9,438,860	928,182	50%	2345
Saskatoon	5,890,710	254,569	49%	2268
Winnipeg	5,306,790	709,253	51%	2353
Toronto	5,905,840	2,809,000	44%	2066
Montreal	4,604,260	1,741,000	44%	2051
St. John's	804,790	111,796	35%	1633
Halifax	5,496,310	414,129	43%	1962
Moncton	2,559,050	72,321	44%	2002

With those numbers, we can calculate the amount of sunlight that these cities receive in a year per square meter.

Total amount of sunlight hitting the Earth x Cloud Cover Efficiency x Hours of Sunlight x (3600 seconds/hour) = Amount of Solar Energy per square meter per year

Vancouver = 1370 W/m^2 x (0.41) x 1938 hours x (3600 seconds/hour) = 3.92 x 10^9 J/m^2

Edmonton = 1370 W/m^2 x (0.50) x 2345 hours x (3600 seconds/hour) = 5.78 x 10^9 J/m^2

Saskatoon = 1370 W/m^2 x (0.49) x 2268 hours x (3600 seconds/hour) = 5.48 x 10^9 J/m^2

Winnipeg = 1370 W/m^2 x (0.51) x 2353 hours x (3600 seconds/hour) = 5.92 x 10^9 J/m^2

Toronto = 1370 W/m^2 x (0.44) x 2066 hours x (3600 seconds/hour) = 4.48 x 10^9 J/m^2

Montreal = 1370 W/m^2 x (0.44) x 2051 hours x (3600 seconds/hour) = 4.45 x 10^9 J/m^2

St. John's = 1370 W/m^2 x (0.35) x 1633 hours x (3600 seconds/hour) = 2.82 x 10^9 J/m^2

Halifax = 1370 W/m^2 x (0.43) x 1962 hours x (3600 seconds/hour) = 4.16 x 10^9 J/m^2

Moncton = 1370 W/m^2 x (0.44) x 2002 hours x (3600 seconds/hour) = 4.34 x 10^9 J/m^2

So on average, these cities receive anywhere from 2.82 to 5.92 x 10^9 J/m^2.

Now that we know that, we need to figure out how much power each city consumes in a year. According to Canadian Geographic, the Royal Canadian Geographical Society, and Shell, the breakdown of energy consumption in each province, per capita (per person), in a given year is as follows:

British Columbia 10,183 kWh/person

Alberta 15,334 kWh/person

Saskatchewan 12,954 kWh/person

Manitoba 11,779 kWh/person

Ontario 11,221 kWh/person

Quebec 11,719 kWh/person

Newfoundland 11,0216 kWh/person

Nova Scotia 10,074 kWh/person

New Brunswick 12,407 kWh/person

These numbers take into account the amount of energy that is used around the house in a given province. This does not take into account transportation or industrial demands for energy. Remember that when we get to our final numbers, we still haven't factored in businesses and industries in these metropolitan areas. Just households.

Calculating the entire household energy consumption per city can be done as follows:

kWh/person x population x (3,600,000 J/kWh)
= Total Household Energy Consumed in a Given City

Vancouver = 10,183 kWh/person x 647,540 people x (3,600,000 J/kWh)
 = 2.37 x 10^{16} J

Edmonton = 15,334 kWh/person x 928,182 people x (3,600,000 J/kWh)
 = 5.12 x 10^{16} J

Saskatoon = 12,954 kWh/person x 254,569 people x (3,600,000 J/kWh)
= 1.19×10^{16} J

Winnipeg = 11,779 kWh/person x 709,253 people x (3,600,000 J/kWh)
= 3.01×10^{16} J

Toronto = 11,221 kWh/person x 2,809,000 people x (3,600,000 J/kWh)
= 1.13×10^{17} J

Montreal = 11,719 kWh/person x 1,741,000 people x (3,600,000 J/kWh)
= 7.35×10^{16} J

St. John's = 11,026 kWh/person x 111,796 people x (3,600,000 J/kWh)
= 4.44×10^{15} J

Halifax = 10,074 kWh/person x 414,129 people x (3,600,000 J/kWh)
= 1.50×10^{16} J

Moncton = 12,407 kWh/person x 72,321 people x (3,600,000 J/kWh)
= 3.23×10^{15} J

On average, these cities consume between 3.23×10^{15} and 1.13×10^{17} J of energy in household consumption.

Now, we bring it all together to find out how much land we would need to cover with solar panels to fill this need.

Energy Used / Energy Produced = Area Needed

Vancouver = (2.37×10^{16} J) / (3.92×10^{9} J/m^2)
= 6,045,918.4 m^2

Edmonton = $(5.12 \times 10^{16}\,\text{J}) / (5.78 \times 10^{9}\,\text{J/m}^2)$
= 8,858,131.5 m^2

Saskatoon = $(1.19 \times 10^{16}\,\text{J}) / (5.48 \times 10^{9}\,\text{J/m}^2)$
= 2,171,532.8 m^2

Winnipeg = $(3.01 \times 10^{16}\,\text{J}) / (5.92 \times 10^{9}\,\text{J/m}^2)$
= 5,084,459.5 m^2

Toronto = $(1.13 \times 10^{17}\,\text{J}) / (4.48 \times 10^{9}\,\text{J/m}^2)$
= 25,223,214.3 m^2

Montreal = $(7.35 \times 10^{16}\,\text{J}) / (4.45 \times 10^{9}\,\text{J/m}^2)$
= 16,516,853.9 m^2
St. John's = $(4.44 \times 10^{15}\,\text{J}) / (2.82 \times 10^{9}\,\text{J/m}^2)$
= 1,574,468.1 m^2

Halifax = $(1.50 \times 10^{16}\,\text{J}) / (4.16 \times 10^{9}\,\text{J/m}^2)$
= 3,605,769.2 m^2

Moncton = $(3.23 \times 10^{15}\,\text{J}) / (4.34 \times 10^{9}\,\text{J/m}^2)$
= 744,239.6 m^2

Looking back at the chart at the beginning of the calculations, we can quickly determine which metropolitan areas stand a fighting chance of being able to provide the necessary area to house the solar panels. These cities are: Edmonton, Saskatoon, Winnipeg, Halifax, and Moncton. As for the other cities, they would need to find additional land or space to house the overflow of their panels or cut down on their energy consumption in order to fit the capacity that they can provide. The reason I say this is because we cannot increase the amount of solar energy that hits the Earth, control the weather, or make the solar panels more efficient (as we already assume that the solar panels are 100% efficient).

Solar energy is increasing in popularity as people desire green energy or even wish to go "off the grid." It is apparent from our calculations that it is possible for a metropolitan area to utilize solar energy to power its households. But there are some key things to take into account, such as the number of sun hours an area gets in a year, the average percentage of sunshine in a day, the metropolitan area's energy consumption rate, the area's climate (does it have mild weather year-round, or does it experience harsh winter conditions?), the city's population, and even the amount of land that is available to the city to house the panels. As we move forward, these are all things that we need to consider as citizens. That's all for today. In the meantime, Keep Looking for the Science in Things.

Put Down The Plastic

(Originally Published July 12, 2018)

We are two weeks into July and two weeks into Plastic-Free Month, or at the very least, single-use plastic-free month. Take a minute and take a look around you. How many plastic items are there? How many will be used once before they are thrown in the bin (hopefully the recycling bin)? How many of those items can be replaced with glass, metal, organic materials (paper, cotton, bamboo, etc.), or even a multi-use plastic substitute? Did you know that the amount of plastic that we produce in one year is equivalent to the weight of every human on Earth combined? (Laville & Taylor, 2017). EVERY YEAR! And around 91% of it isn't recycled (Geyer et al., 2017). So where does it all go, and what happens when it gets there? Today's post will discuss that and more.

The Making Of

Plastic is made up of various materials. Some are derived from plant resins, such as natural rubber, which comes from the sap of the rubber tree (Professor Plastics, 2011). However, most plastic materials that you find on the market today are made up of natural gas, oil, or coal, otherwise known as hydrocarbons (Professor Plastics, 2011). In a given year, we use roughly 1.6 million barrels of oil to make all the disposable water bottles we use (Leblanc, n.d.). The exact process of making plastic is quite complicated and varies depending on the type being made (e.g. polyethylene, polystyrene, or polypropylene), but the simple version of the procedure is as follows: Hydrocarbon molecules are rearranged and strung together to create chains of molecules (called polymers) (Professor Plastics, 2011). These polymers are then assembled to form sheets and pellets that are then shaped and moulded into various objects (Professor Plastics, 2011). What makes plastic so desirable over other materials is the polymers' lightweight, high strength-to-weight ratio, and their ability to be moulded into almost any shape imaginable.

Disposal and Breakdown

Unfortunately, because of the way the molecules are arranged and strung together synthetically, they take a long time to break down naturally, if at all. On average, it takes plastic objects around 1000 years to break down in landfills, with plastic bags taking 10 to 1000 years and disposable water bottles taking 450 plus years to break down (Leblanc, n.d.). And even though you can no longer see it, it is still around as toxic chemicals that are leaching into the soil and poisoning the environment ("What's the problem with plastic bags?" 2014).

Recycling

So what happens when you recycle your plastic? Hopefully, many of you, if not all, are recycling. Well, after you thoroughly wash your recyclables and put them either in recycling bags to be picked up from your curb or you take them to a recycling depot, they are transported to a recycling facility where they are sorted (either by type, colour, or how they were made) ("How Is Plastic Recycled," 2018).

Next, the plastic is washed to ensure all food bits, labels, and dirt are removed from the items, which can't be recycled ("How Is Plastic Recycled," 2018). Any impurities can impact the reforming process and result in the new product not forming properly (which could result in it having to be thrown out) ("How Is Plastic Recycled," 2018). This process goes a lot quicker when everyone washes their recyclables, so do them a favour and scrub, scrub, scrub.

Now that it is clean, it is put through an industrial shredder, which breaks the plastic down into small pieces ("How Is Plastic Recycled," 2018). After this step, all the bits pass through either a metal detector or a magnet to ensure that there is no metal in the mix ("How Is Plastic Recycled," 2018).

Once it is deemed that only plastic bits remain, it goes through an identification and separation process, which determines the quality and class ("How Is Plastic Recycled," 2018). First, sorters check the plastics' density by floating the pieces in a large water tank ("How Is Plastic Recycled," 2018). Denser plastics sink, while less dense ones float. From here, the pieces are placed in a small wind tunnel to determine the thickness of each piece ("How Is Plastic Recycled," 2018). Much like in the water tank, smaller pieces are near the top of the wind column, while thicker pieces are closer to the bottom. Finally, the pieces are sorted out by melting point and, if not already, by colour ("How Is Plastic Recycled," 2018).

Finally, once it is completely sorted out, it is compounded. During this process, the small bits are melted together to form pellets, which are then sent to the manufacturer, which turns them into something new ("How Is Plastic Recycled," 2018).

Despite how intensive the recycling process is, it is still better than having to acquire more natural gas, oil, or coal out of the ground to produce more while the discarded pieces remain in landfills for centuries.

Facts & Figures

One of the things I find fascinating is the facts and figures that surround plastic, so I thought I would share some notable ones here:

- 160,000 plastic bags are produced every second; in a year, that is enough to circle the Earth seven times ("What's the problem with plastic bags," 2014).

- 500 million plastic drinking straws are used in the United States every day (Parker, 2017).

- 8 million metric tons of plastic are introduced to the oceans every year (Geyer et al., 2017).

- The Great Pacific Garbage Patch, which is mostly composed of plastic, is twice the size of Texas ("What's the problem with plastic bags," 2014).

- In the North Pacific Ocean, there is six times more plastic than there is plankton ("What's the problem with plastic bags," 2014).

- One million seabirds and 100,000 marine mammals are killed every year by plastic in the ocean ("22 Facts…" 2014).

- Their chemicals can be absorbed by the body, some of which alter hormones or cause health problems ("22 Facts…" 2014).

- According to the UN, there are more microscopic plastic particles in the oceans than there are stars in the Milky Way galaxy ("Turn the tide on plastic," 2017).

- If we don't reduce the amount we throw out, plastic pollution in the oceans will outweigh the entire biomass of fish by 2050 ("The New Plastics Economy," 2016).

How You Can Help

Now I know we all care about the planet and don't want to be overrun by plastic pollution, so there are some things that you can do to help out:

- Recycle, Recycle, RECYCLE. Whether you take it to the recycling depot or repurpose it somewhere in your home, single-use containers can always be repurposed or recycled and given a new life (I use the tote from my laundry detergent to store things such as batteries to be recycled, sharps to be recycled, etc.).

- Go out and pick up garbage. Whether it is in the streams, ditches, or on the beach, it all has the potential to harm the environment.

- Avoid plastic when possible. Use reusable grocery bags, beeswax food storage wraps, etc. Also, if given the option of paper or plastic, choose paper. Order online? Ask sellers to package your goods in cardboard boxes wrapped in newspaper or cardboard cushioning. And support companies that use less packaging.

- Buy in bulk. In Canada, we have an amazing store called Bulk Barn, where you can buy all your dry pantry goods in bulk. They even encourage you to bring in your reusable containers to take home your groceries.

- When you can't avoid it, opt for multi-use plastic that is BPA-free.

- Say goodbye to disposable straws. Starbucks plans to no longer carry them by 2020. And many cities are jumping on board too. There are plenty of options, too. From multi-use plastic straws to stainless steel, glass, silicone, and paper, one restaurant has even opted for pasta straws.

At the end of the day, we can all do our part to minimise the impact of plastic and its pollution on the planet. It's our responsibility to do so. In the meantime, Keep Looking For the Science in Things.

<u>Footnote:</u>

Since the original blog post was written in 2018, I am happy to report that more countries have jumped on the bandwagon of ditching single-use plastics. The following statistics come from reports released by the United Nations.

As of December 2018, 127 out of the 192 countries reviewed by the UN have national legislation on plastic bags. 83 of these countries have adopted a ban on free retail distribution. 61 of these countries have a manufacturing and importation ban on plastic bags.

For more details on the UN report on "Legal Limits on Single-Use Plastics and Microplastics: A Global Review of National Laws and Regulations," see the included citation in the appendix.

Breath of Fresh Air

(Originally Published August 16, 2018)

Wildfires are roaring across much of western North America. British Columbia has declared a province-wide state of emergency. Whether you are on evacuation alert or a couple of provinces or states away from the fires, the fact that these fires are occurring is inescapable because the smoke is EVERYWHERE. With all the smoke in the air, it affects the air quality, and with poor air quality, everyone suffers. In today's post, we will be discussing air quality and how it's measured.

Gauging Air Quality

Air quality can be altered by various forms of air pollution, which include dust, emissions (such as from vehicles, factories, etc.), sandstorms, and smoke. In turn, they can be influenced by a change in wind direction, lack or presence of wind, temperature, and other changes in weather. Poor air quality not only results in limited visibility and changes in local weather but can have adverse effects on our health. A subset of the population faces greater risks from poor air quality: people with diabetes, people with lung disease or respiratory problems, seniors, pregnant women, children, and individuals participating in sports or strenuous work outdoors ("Air Quality," n.d.).

Air Quality Index/Air Quality Health Index

Fortunately for us, air quality is monitored as a part of most meteorological and weather monitoring organisations and is classified on two different scales: the Air Quality Index and the Air Quality Health Index.

The Air Quality Index (AQI) is a scale that ranges from 0 to 300+ (The World Air Quality Project, n.d.). Its purpose is to indicate the specific level of an individual pollutant. The number that you see for the AQI value is the level of the greatest single pollutant. It does not account for the other pollutants present. This scale is useful for determining the risks that are posed to both humans and the environment. There are six groups of AQI values, and they are as follows:

- AQI 0 – 50: Good Air Quality. Air pollution presents little to no risk.

- AQI 51 – 100: Moderate. Air quality is acceptable, but depending on the pollutant, there may be health concerns for those at risk.

- AQI 101 – 150: Unhealthy for Sensitive Groups. Those who are at risk will experience health effects. The general population will be relatively unaffected.

- AQI 151 – 200: Unhealthy. Everyone will begin to experience health effects; those at risk will experience more serious health effects.

- AQI 201 – 300: Very Unhealthy. Emergency conditions. The entire population is likely to be affected in one way or another.

- AQI 300+: Hazardous. Health Alert. Everyone may experience serious health effects.

The Air Quality Health Index (AQHI) is a scale that ranges from 1 to 10+ ("How to use the Air Quality Health Index," 2016). Its sole purpose is to indicate the risk that air quality presents to human health from a combination of pollutants. This system is commonly used in Canada because, rather than having to isolate individual air pollutants, it takes an overall assessment of air quality and how it relates to human health. For those who are not familiar with the scale, it goes as follows:

- AQHI 1 – 3: Presents low risk to the public, including those that are at higher risk. Air quality between 1 and 3 is ideal, and no safety or health measures need to be taken.

- AQHI 4 – 6: Presents moderate risk to the public. Those who are at risk should take precautions (depending on their sensitivity). As for the general population, they do not need to make changes to their daily activities, unless they begin to experience symptoms (e.g. coughing, throat irritation, or difficulty breathing).

- AQHI 7 – 10: Presents a high risk to the public. Those who are at risk should take precautions and reduce or reschedule outdoor activities. As for the general population, those who experience symptoms should consider minimising or eliminating outdoor activities.

- AQHI 10+: Presents a very high risk to the public. Those who are at risk should take all necessary precautions and avoid any strenuous activity outdoors. Those who are at higher risk should minimize their time spent outdoors. As for the general population, any demanding work outdoors should be reduced or rescheduled for a day with better air quality.

How is The Air Quality Health Index Determined?

Because the Air Quality Health Index is an indicator of overall air quality and how it relates to human health, it uses three main pollutants rather than one ("Air Quality," n.d.). These are ground-level ozone, fine particulate matter, and nitrogen dioxide. Ground-level ozone is made up of a naturally occurring gas, however, due to human activities, we have increased the amount that is present in our atmosphere. This overabundance can especially be seen in the summertime, as it contributes to the formation of smog and is a contributing factor in large-area air pollution episodes (e.g. Los Angeles). Next, fine particulate matter. Anyone downwind of the wildfires is experiencing high levels of fine particulate matter in their air. It can also be caused by vehicle emissions, industrial emissions, and other sources. Fine particulate matter can affect local areas or be more widespread as it is carried downwind by jet streams. Finally, nitrous dioxide is an air pollutant that is emitted by vehicle emissions and facilities that rely on fossil fuels. Not only is it an air pollutant in its own right, but it also aids in producing ground-level ozone and fine particulate matter, thus potentially amplifying an area's level of air pollution. It is commonly seen in areas of high traffic or regions of high industrial activity.

Air Pollution Safety

As air quality diminishes, or as many are experiencing poor air quality right now, you may want to take measures to protect you and your family's respiratory health. Here's a list of things you can do:

- Ensure that the filters in your homes and vehicles are regularly changed.

- If you are particularly sensitive to air quality, you may want to invest in an air purifier for your home and workspace.

- If you work outside, you may want to minimize time spent outside or use a face mask or respirator when you spend an extended period of time outside.

- Use a face mask or respirator when outside.

- Keep windows and doors closed.

- Keep your pets inside and monitor them when they are outside. Your furry friends are particularly sensitive to poor air quality, and they may not know why they are having trouble breathing. This can make them scared or on edge, so they may be out of sorts.

Air quality is vital to our environment and our health. Poor air quality can pose a challenge, and sometimes there isn't much that we can do to improve it, such as during wildfires or sandstorms. However, there are temporary measures that we take to improve our conditions and prevent harmful effects on our health. That's all from us for today. In the meantime, Keep Looking for the Science in Things.

The ABCs of Climate Science

(Originally Published November 7, 2018)

To date, one of the most popular posts we've released is Schrodinger's Cat's Quick Guide to Climate Change. And we couldn't be more thrilled about all the interest in this very important issue that we are facing (I personally am excited about all the interest because climate change science is an area of science that I have fallen in love with; total climatology nerd here). But one thing is for sure: there is a lot of jargon and technical terms used when scientists talk about climate change, and a lot of the time, this makes climate change go over the general public's head and often leads to the question, "Why should I even care?" Well, Schrodinger's Cat has come to the rescue. In today's post, we will be breaking down all the most common and complex terms in the ABCs of Climate Science.

Albedo – the amount of solar energy that is reflected by a surface or object. We all know that the lighter the colour of the object, the more solar energy is reflected by that object. So surfaces such as snow, ice, and light-coloured sand have a higher albedo than the ocean, forested regions, and asphalt. As you may have noticed, in recent years, more and more developers are opting for painting or tinting their concrete to be lighter in colour. Another thing that affects the Earth's albedo is the amount of cloud cover in the region.

Business As Usual – you may have heard different countries' governments say that they are adopting a "Business As Usual" approach or policy towards climate change. Bad. Bad. BAD!!! By stating this, they are stating that they are not taking on any new actions or plans to help fix the problem of climate change. So in other words, they will not be reducing the amount of emissions that they are emitting.

Coriolis Effect – I know you've heard the stories about how when you flush a toilet in the Northern Hemisphere, it rotates one way, and if you flush a toilet in the Southern Hemisphere, it rotates the other way.

Sorry to burst your bubble, but that isn't true. But it is based on the concept of the Coriolis effect. This phenomenon is caused by the Earth's rotation and affects major weather systems, mainly cyclones. You may have noticed that cyclones in the Northern Hemisphere tend to rotate counterclockwise, and clockwise in the Southern Hemisphere. Coriolis is to blame. So why doesn't it apply to toilets? Toilets are too small of a system for Coriolis to take effect. However, if there was a whirlpool out in the ocean, Coriolis would apply in that case.

Diabatic Process – For those familiar with physics, chemistry, and thermodynamics, you may be familiar with diabatic processes. Diabatic processes are changes to the system during which the system exchanges energy with its surroundings via a temperature gradient. One example of a diabatic process in relation to climate change is that a lot of the excess heat in the atmosphere is transferred to the cooler ocean, resulting in the ocean warming up and the atmosphere seemingly unchanged. Because of the diabatic process, you'll commonly hear, "If it's supposed to be global warming, why isn't it warming outside?" This is why. The ocean is compensating for the increase in atmospheric temperature. Not good. Poached fish, anyone?

El Nino Southern Oscillation (ENSO) – is a phenomenon that occurs in the Indian and Southern Pacific Oceans. You may be familiar with El Nino, which is a warm water current that flows along the coast of Ecuador and Peru, as a result of trade winds weakening, causing warm water to flow across the surface from Indonesia to South America and causing downwelling of that nutrient-rich water, hurting the fishing industry. This event occurs every two to seven years and lasts anywhere from nine months to two years. While this is happening, there is also a change in atmospheric pressure patterns and circulation, creating a change in weather patterns, which is known to affect the weather even in North America. Together, they make up the El Nino Southern Oscillation, or ENSO. Opposite El Nino is La Nina, which is a strengthening of the trade winds, so cold water flows across the surface from South America to Indonesia. As a result of this, during La Nina, there is a strong upwelling of warm water off the coast of Peru and Ecuador, which causes the fishery industry on the West coast of South America to flourish.

Feedback Loop – If you venture into the world of climate change, feedback loops become everything. They can be either positive, in which they continue to add to the system, or negative, in which they continue to take away from it. Also, they can be runaway feedback loops, where they go unchecked and their effect compounds and builds, while some feedback loops trigger other feedback loops, which help keep the system constant. On Earth, we have several feedback loops that maintain a habitable constant. In climate change, the feedback loops of concern are in relation to warming. Albedo contributes to many feedback loops. As ice caps melt, they become smaller and smaller (obviously), which creates a smaller and smaller area that reflects the Sun's energy back and a larger and larger area that absorbs the Sun's energy (ocean or land), which increases that surface's temperature.

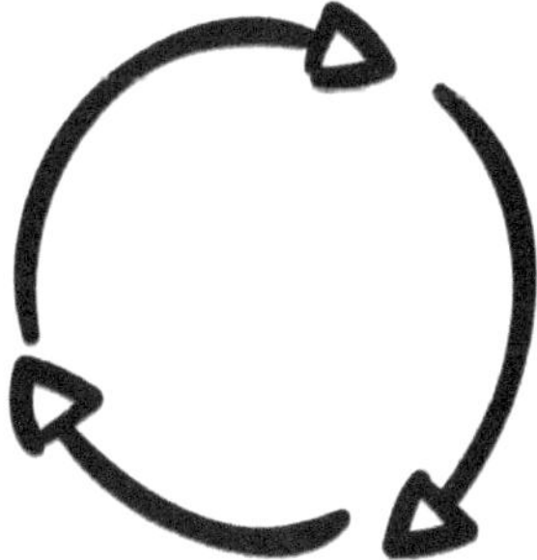

Greenhouse Gases – We have previously discussed greenhouse gases in our Quick Guide to Climate Change. But it doesn't hurt to repeat ourselves. Greenhouse gases can be either natural (ozone, methane, nitrous oxide, and carbon emissions) or human-made (CFCs, anthropogenic carbon emissions, perfluorocarbons, hydrofluorocarbons, sulphur hexafluoride). These gases trap heat emitted from Earth, creating the greenhouse effect.

Holocene – We are currently in the geological era of the Holocene, which started about 11,600 years ago, at the conclusion of the last Ice Age. It's probably not as exciting as other eras (we don't have dinosaurs), but it has been an era of dramatic changes when it comes to climate.

Ice Cores – Sorry, this is my climate science geek coming out here. Ice cores are just the coolest. Most ice cores that scientists study come from regions that are permanently frozen (the Arctic and Antarctic regions), so there are hundreds and thousands of years of layers of ice built up. They allow scientists to analyse the past climate as well as major global events. If there is a deposit of ash in a certain area, scientists can then speculate that there was a great forest fire or a major volcanic eruption. In fact, looking at the layers, you can determine the start of the Industrial Revolution, as it was the beginning of large amounts of human-caused air pollution. If you find yourself in the Edmonton area, be sure to check out TWOSE, the Telus World of Science, as they have a great display on ice cores in their Environment Gallery. I practically run to it every time I get there.

Keeling Curve – Odds are, if you follow any climate change news, you have seen a Keeling Curve. It's a graph that depicts the change in the concentration of atmospheric carbon dioxide. It is based on the measurements that have been collected at the Mauna Loa Observatory in Hawaii since the 1950s. You may notice that the overall trend is in an upward motion, which means that since the 1950s, the amount of carbon dioxide in Earth's atmosphere has increased. However, the line moves up and down each year. This is due to the change of seasons. In Spring and Summer, plants uptake carbon dioxide from the atmosphere as they produce oxygen. In Fall and Winter, the plants die off and are unable to uptake carbon dioxide, so the level increases until next Spring when the plants begin to uptake carbon dioxide once again.

Latent Heat Flux – Latent heat flux is the flow of heat from the Earth's surface to the atmosphere that is associated with the evaporation or condensation of water vapour at the surface. Evaporation involves the taking in of heat energy, whereas condensation involves the emission or removal of heat energy. So as water droplets heat up, they evaporate into the atmosphere, where the water vapour cools down, emitting heat energy into the atmosphere as it falls back to the Earth's surface as a water droplet.

Mauna Loa – Located in picturesque Hawaii, the Mauna Loa Observatory is located on the famous Mauna Loa volcano. This observatory experiences some of the most ideal conditions to not only make astronomical observations but also climate observations, as it has some of the cleanest air quality in the World. From here, we get our measurement for the Keeling Curve, which measures the concentration of atmospheric carbon dioxide since the 1950s.

Natural Greenhouse Effect – Not all of the greenhouse effect is bad; without it, the Earth's average temperature would be below freezing. There would be no liquid freshwater, and life for humans would be rather difficult, if not impossible. Some greenhouse gases are natural (humans didn't emit them into the environment), such as ozone, carbon dioxide, and water vapour, and they allow the Earth to be habitable.

Ocean Acidification – Not only does the ocean absorb a lot of atmospheric heat, but it also absorbs around 25% of all human-made carbon dioxide that is in the atmosphere. This helps reduce the amount that it affects our climate. But not all of the carbon dioxide that is absorbed by the ocean is converted to oxygen by phytoplankton. A good amount of carbon dioxide dissolves in seawater and becomes carbonic acid, which makes the water more acidic. As a result, it hinders marine animals' ability to build their shells or skeletal structures, kills corals, and breaks down calcified habitats, which impacts fish populations.

Permafrost – With climate change, we often hear about the thawing of the permafrost. But what is permafrost? Permafrost is frozen ground that remains frozen (below 0°C) throughout the year for many years. By thawing, not only does it mean that our planet is getting warmer, but large areas of permafrost are actually frozen marshes and swamps, which hold carbon dioxide. When they thaw, they release this carbon dioxide into the atmosphere, adding to our problem.

Radiative Forcing – Radiative forcing is the change in the difference between the incoming solar radiation and the emitted infrared radiation. Without radiative forcing, the amount of energy coming into our atmosphere would be the same as the amount of energy leaving. However, thanks to greenhouse gases, we are able to trap some of the energy trying to leave (this is called positive radiative forcing, as incoming energy is greater than what is leaving). As we add more and more greenhouse gases to our atmosphere, we increase the amount of radiative forcing that we are experiencing on Earth.

Sea Ice Minimum/Maximum – The sea ice minimums and
maximums are measured both in the Arctic and Antarctic; however, the
Arctic is the most commonly shown. The sea ice minimum reflects the
total amount of sea ice present at the end of Summer (usually in
September). This reflects how much ice has melted since Winter. The sea
ice maximum reflects the total amount of sea ice present at the end of
Winter, after all that could have frozen has frozen. The reason climate
scientists monitor these so closely, the warmer our planet gets, the less
ice there is at both the minimum and the maximum, the less habitat that
is available for the animals that rely on the ice to hunt, travel, and live.

Transpiration – The process by which water vapour is evaporated
from the surfaces of leaves through pores called stomata. The more
transpiration that occurs, the more water vapour, a greenhouse gas, is
present in the atmosphere, amplifying the greenhouse gas effect.

Urban Heat Island – Ever notice that when you are in the city
during the Summer, it can be so hot, especially compared to the
surrounding rural area? This is due to many factors that result in an
urban heat island. These factors include the increased amount of runoff
(as rainwater cannot penetrate concrete); the heat retention of the
concrete, steel, etc.; the lower surface albedo value as darker materials
hold the Sun's energy rather than reflecting it back; and the increased
concentration of air pollution.

Vulnerability – With climate change occurring so rapidly, it can be
difficult to begin to think of what regions need our immediate
attention. That is where scientists assess the vulnerability of the area.
Vulnerability is determined by the characteristics of the area (e.g.
low-lying coastal regions, permafrost, delta, etc.), how significantly the

area is being affected (e.g. increased flooding, sinkholes), the rate at which climate change is occurring there (e.g. the Arctic and Antarctic poles may see a more drastic change than a more temperate region), the region's sensitivity to the change (e.g. coral reefs are extremely sensitive to changes in temperature and ocean acidity compared to kelp forests); and the region's ability to adapt (e.g. A lot of wetlands, such as bayous, are highly adaptable as they are brackish waters, meaning that they are a combination saltwater and freshwater, so as ocean levels rise, the region's ecosystems will adapt accordingly. Polar regions are less adaptive, as are many of the low-lying tropical islands that will be swallowed up by the ocean).

Weather – Weather is different from climate, as it is a temporary system that eventually changes. Climate is a more long-term condition. Climate can impact weather, but weather does not impact climate.

Climate change is a complex issue that the modern World faces. A lot of confusion surrounding climate change stems from complex terms that are being thrown around by experts. Hopefully, today's post has shed some light on climate change so you can better understand future climate change reports. Knowledge is power, and with a better understanding, we can reduce the impacts of climate change. That's all from us for today. In the meantime, Keep Looking for the Science in Things.

Health Science

Unraveling Your DNA
(Originally Published April 25, 2018)

I'm sure everyone has seen the commercials on television where the guy is on there stating he was confused when he didn't see many Germans in his family tree, having grown up German, and that he took an at-home DNA test and found out he was more from the British Isles, so he "traded in his lederhosen for a kilt," or the woman that got the shocking results back that she isn't Hispanic, but rather she's "from all nations." These commercials and companies that conduct at-home DNA tests are everywhere, from assessing your ancestry to even looking at your health and what conditions you may get based on your DNA. But if you are like most people, you have some questions about these tests. How do they work? Do they even work? How accurate are they? If my results say that I have the gene for medical condition "A," does that mean I'm going to get it? Well, today, we are going to look into all these questions and more.

First of all, let's brush up on a little DNA and genetics knowledge. As many of us know, DNA is the genetic material that makes us who we are, and it's assembled into a code that is unique for each person (NIH, n.d.). This unique code not only distinguishes your gender and how you look (hair colour, skin tone, eye colour, etc.), but it also contributes to any medical conditions that you have (likelihood to develop high blood pressure, diabetes, genetic diseases such as Huntington's disease, MS, etc.). This important information is carried in proteins called chromosomes, which are found in all our cells (except our red blood cells and specific cells in our skin, hair, and nails (cornified cells)). Humans have 23 pairs of chromosomes, for a total of 46 chromosomes. One-half of these pairs came from your mother, and the other half came from your father (in the future, it may come from your mom and mom or your dad and dad). I'm not going to go into exactly how that happens (for further information, refer to the birds and the bees discussion you had in junior high). So essentially, half of your DNA came from your mom and the other half from your dad (other than mutations, but that is a whole different can of worms). And they got their DNA code from their parents, all the way back through history. So in addition to having half of each of your biological parents' DNA, you also have one-quarter of each of your grandparents' DNA, and so on.

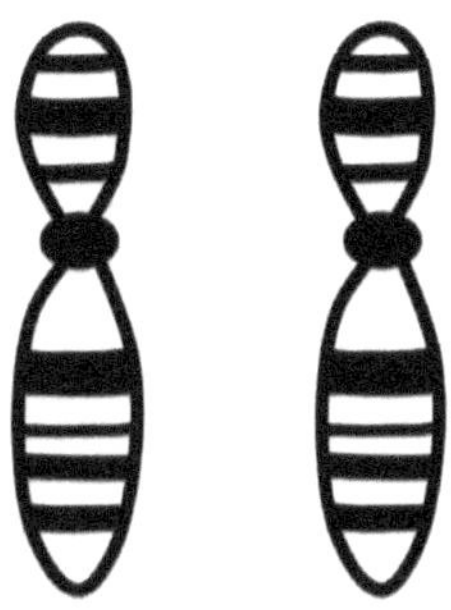

How do they work?

Now scientists have identified certain markers in people's DNA codes that identify certain regions and groupings of people. That is how, theoretically, these at-home DNA tests work. Their labs look for these markers that identify a region or group of people, and the greater the number of these markers that identify a region, the greater the percent that you are from there. And you are matched with relatives based on the similarity of your codes to each other. As for the medical component of some of these tests, certain portions of the human genetic code have been identified as markers for medical conditions; if these markers appear in your DNA, you have a chance of getting the condition. It is similar to when your doctor asks for your medical history as well as that of your parents and grandparents. If you have family members that have a medical condition, there is a chance that you have genetically inherited it from them as well. But how do these at-home kits get your DNA? Well, there are many ways that they collect it; it depends on the test ("How it Works," n.d.). You purchase the kits on the company's website, and then they mail you a kit in order to collect your DNA sample. Some tests require you to spit in a vial and mail the vial back to them.

Others require swabs from the inside of your mouth. Those are the less invasive or less personal ways of getting your DNA; other companies ask for blood samples, while others ask for a sample of your, ahem, poop. All these samples are shipped in by you to the company's lab, where the samples are analyzed, and then they send you back your results.

Do they even work?

When it comes to determining your heritage as well as that of potential other relatives through your DNA, the major names in at-home testing (23andMe, AncestryDNA, and MyHeritage) are quite accurate, with many news broadcasters conducting experiments to determine accuracy (Rossen and Bomnin, 2017). However, the tests to identify the potential for medical conditions as well as characteristics are not as accurate, with many tests claiming that they are not actual diagnostic tests and many doctors stating that if you have concerns, you should discuss those with your doctor and even take your results to your doctor (Ellen, 2017 | Hatfield, 2007).

If a medical condition is identified, does that mean I am automatically going to get it?

Not necessarily (Park, 2017). Some conditions can be prevented, such as high blood pressure and type two diabetes, through diet and exercise, and that's where these tests can be useful, so you can be proactive in treating them before they even occur. Also, other ailments are what we call recessive, which means that they are the non-dominant version of that segment of the genetic code. So unless you have that segment of code on both chromosome pairs that holds that information, you may never experience any symptoms or expression of that ailment, and your children may never experience the symptoms or express it, or even be carriers of it. So what is Schrodinger's Cat's opinion on these tests? If you want to learn more about your heritage and possibly uncover some relatives that you didn't know about, do your homework and then read the reviews of the major names in DNA heritage testing (such as Ancestry, 23andMe, My Heritage, etc.) and pick which one you think would be best for you when it comes to the information that you will get back as well as the means of collecting your DNA. But when it comes to the medical portion of these DNA tests, you are better off to go see your doctor if you have any concerns or schedule an appointment with a geneticist, as they can actually sit down with you and discuss everything and answer any concerns you may have. DNA analysis is still a budding field of study, and there are many aspects of it that we just don't know, so when it's something as crucial as your health, it's best to leave it to the experts, at least for now.

Thank you for tuning into today's post on Schrodinger's Cat. And remember Keep Looking for the Science of Things.

Can Sunscreen Really Give You Cancer?

(Originally Published May 23, 2018)

Ok, this is one that you may or may not have seen floating around social media, and this may be one of the more dangerous thoughts floating around right now as we head into Summer and sunny weather. We all know that the UV rays from the Sun can cause melanoma and skin cancer. There have been countless studies done to show the effects of overexposure to UV rays and the Sun on the skin. But what about sunscreen? Are there chemicals in sunscreen that also put us at risk for cancer? Let's take a look.

There are many different brands and types of sunscreen on the market today. But after a trip to the local drug store, I have compiled a general ingredients list from over 10 different sunscreens (ingredients in bold are in 3 or more different types of sunscreen):

Active Medical Ingredients: Avobenzone, Homosalate, Octisalate, Octocrylene, Oxybenzone, Titanium Dioxide, and Zinc Oxide

Inactive Ingredients: Acrylates/C10-30 Alkyl Acrylate Crosspolymer, Acrylates/C12-22 Alkyl Methacrylate Copolymer, Acrylates/Dimethicone copolymer, Acrylates/Octylacrylamide copolymer, Alcohol Denat., Alkyl Acrylate Crosspolymer, Aloe Barbadensis Leaf Juice, Alumina, Aluminum Starch Octenylsuccinate, Ammonium Acryloyldimethktaurate/VP Copolymer, Aqua (Water), Ascorbyl Palmitate, Avena Sativa (Oat) Kernel Extract, Avena Sativa

(Oat) Kernel Flour, Behenyl alcohol, Benzyl alcohol, BHT, Blue-1, Butylene Glycol, Butyloctyl Salicylate, Butyrospermum Parkii (Shea) Butter, Caprylic/Capric triglyceride, Caprylyl Glycol, Caprylyl Methicone, Carbomer, Carica Papaya (Papaya) Fruit Extract, Cellulose gum, Cera Alba, Cetearyl alcohol, Cetearyl glucoside, Ceteth-10 phosphate, Cetyl alcohol, Cetyl dimethicone, Cetyl dimethicone/Bis-Vinyldimethicone Crosspolymer, Cetyl PEG/PPG-10/1-Dimethicone, Chlorphenesin, Chrysanthemum Parthenium (Feverfew) Flower/Leaf/Stem Juice, Codium Tomentosum Extract, Colocasia Antiquorum Root Extract, Coco-Glucoside, Cyclopentasiloxane, C10-18 triglyceride, C12-15 Alkyl Benzoate, C12-22 Alkyl Methacrylate Copolymer, C-30-38 Olefin/Isopropyl Maleate/MA Copolymer, Dicaprylyl Carbonate, Dicetyl phosphate, Diethylhexyl syringylindenemalonate, Diethylhexyl-2, Diethylhexyl-2,6-Naphthalate, Diisopropyl Adipate, Dimer Dilinleate Copolymer, Dimethicone crosspolymer, Dipotassium glycyrrhizate, Disodium EDTA, Disteardimonium Hectorite, Eicosene Copolymer, Ethylhexyl Benzoate, Ethylhexylgylercin, Ethylhexyl Methoxycrylene, Ethyl Methicone, Eucalyptus Globulus Leaf Extract, Gamma-Octalactone (Coconut), Gamma-Undecalactone (Sweet Fruit), Glycerin, Glyceryl Behenate, Glyceryl isostearate, Glyceryl stearate, Hexyl Laurate, Hydrogenated Methyl Abietate, Hydrolyzed Hyaluronic Acid, Hydrolyzed Oat Protein, Hydroxyacetophenone, Isohexadecane, Isostearate, Jojoba Alcohol, Lauryl alcohol, Lauryl PEG-8 Dimethicone, Lecithin, Mangifera Indica (Mango) Seed Butter, Methyl Lactate, Methylisothiazolinone, Methylparaben, Mica, Microcrystalline cellulose, Myristyl alcohol, Neopentyl Glycol Dilheptanoate, Octylacrylamide copolymer, Octyldodecyl citrate crosspolymer, Palmitic acid, Panthenol, Paraffin, Parfum (Fragrance), Passiflora Incarnata Fruit Extract, PEG-8, PEG-8 Laurate, PEG-10 Dimethicone, PEG-100 Stearate, Pentylene Glycol, Phenethyl Alcohol, Phenethyl Benzoate, Phenoxyethanol, Phenylisopropyl Dimethicone, Phenyl Trimethicone, Plumeria Acutiflora Flower Extract, PPG-5-Ceceth-20, Polyester-8, Polyglyceryl-3 Sterate/Isostearate/Dimer Dilinleate Copolymer, Polygyceryl-4 Isostearate, Polyhydroxystearic acid, Polymethylsilsesquioxane, Polyurethane-62, Porphyra Umbilicalis (Red Algae) Extract, Potassium Jojobate, Potassium Palmitoyl Hydrolyzed Oat Protein, Propanediol, Propylene Glycol 2, Propylene Glycol Dibenzoate, Propylparaben, Psidium Guajava Fruit Extract, Retinyl Palmitate, SD alcohol 40-B, Silica, Simmondsia Chinensis (Jojoba) Butter, Sodium Acryloyldimethyltaurate/VP Crosspolymer, Sodium Ascorbyl Phosphate, Sodium Citrate, Sodium Chloride, Sodium Gluconate, Sodium Hydroxide, Sodium Polyacrylate, Squalane, Stearic acid, Stearyl alcohol, Styrene/Acrylates Copolymer, Synthetic Wax, Terminalia Ferdinandiana (Kakadu Plum) Fruit Extract, Tocopherol, Tocopheryl Acetate, Trideceth-6, Triethanolamine, Triethyl citrate, Trisiloxane, Violet-2, VP/Eicosene Copolymer, Xanthan Gum, 6-Naphthalate

If you didn't read the entire list, I don't entirely blame you. In today's post, we are only going to discuss the active ingredients. But I would like to put emphasis on the importance of reading the labels of the products that you buy. You might surprise yourself, in a good way and a not-so-good way. Now, I know there won't be many expert chemists that will be reading today's post (I am certainly not an expert chemist), and researching the individual ingredients is time-consuming while standing in the pharmacy aisle, so there are some apps for your phone that can help you out. One of my favourites is the Think Dirty. App. It's very easy to use. You scan the barcode of the product you are looking at, and it lists all the dirty ingredients, half-and-half ingredients, and clean ingredients found in the product, as well as an overall score for the product and their clean recommendations. I surprised myself with what products I will not be buying again in the future.

Now, let's take a look at sunscreen's active ingredients:

Avobenzone – Chemical sunscreen, absorbs UV rays and converts them into energy that is less damaging to the skin (MacLeman, "Avobenzone," n.d.). Toxic when combined with certain ingredients and if used too often or in large doses. Standard sunscreen concentrations (3-5%) are safe for human use. There have been no reported negative effects since FDA approval in 1998. Minor skin allergies have occurred.

Avobenzone

Homosalate – Chemical sunscreen, absorbs UV rays ("Homosalate," n.d.). Possible endocrine disruptor, specifically estrogen, androgen, and progesterone production (Schreurs et al., 2002 | Ma et al., 2003 | Krause, 2012). Some studies have shown that when introduced to human breast cancer cells, cell growth increases (Jiménez-Díaz et al., 2013). Also not recommended for use by pregnant and nursing mothers, as it has been found in placental tissue, as well as human breast milk (Jiménez-Díaz et al., 2013 | Schlumpf et al., 2010). Use of homosalate makes an individual more susceptible to skin absorption of herbicides and pesticides when worn with bug spray containing DEET (Pont et al., 2004 | Brand et al., 2003) .

Homosalate

Octisalate – Chemical sunscreen that stabilizes avobenzone and prolongs UV protection ("Octisalate," n.d.). Possible allergic reactions can happen. Enhances chemicals' ability to be absorbed into the skin (Walters et al., 1997). Possibly toxic to the environment (Layton, 2015).

Octisalate

Octocrylene – Chemical sunscreen, neutralizes UV rays and increases the skin protective coating of UV protectors (MacLeman, "Octocrylene," n.d.). Easily absorbed through the skin, thus increasing the likelihood of free radicals and bad chemical reactions occurring under the skin. Harmful in large doses. Not advised for use by women who are pregnant, as reproductive toxicity is possible.

Octocrylene

Oxybenzone – Chemical sunscreen, protects from UV damage. Oxybenzone belongs to the chemical family Benzophenone, which are persistent (difficult to get rid of), bioaccumulative (builds up in your body over time), and toxic, or PBT ("Benzophenone & Related Compounds," n.d. | Kim & Choi, 2014). They are a possible carcinogen (cancer-causing agent) and endocrine disrupter; however, this is debatable (DiNardo & Downs, 2019 | World Health Organization, n.d. | Morison & Wang, n.d.). Also, it could cause developmental and reproductive toxicity, organ system toxicity, irritation and potentially toxic effects on the environment (Weisbrod et al., 2007 | Fediuk et al., 2011 | Agin et al., 2008 | Kim & Choi, 2014).

Oxybenzone

Titanium Dioxide – Mineral sunscreen, UV filter, naturally occurring powder ("Titanium Dioxide," n.d.). When in lotions and creams, very low risk, as it is unable to penetrate through healthy skin ("Opinion on Safety of Nanomaterials in Cosmetic Products," n.d.). It only poses a cancer risk when it is directly inhaled. No other health concerns.

Zinc Oxide – Mineral sunscreen, creates a protective barrier layer on the skin against UV rays (MacLeman, "How Does Zinc Oxide Benefit Your Skin?" n.d.). It is deemed anti-allergenic and doesn't clog pores; however, some allergic reactions have been recorded, although they are uncommon. Can build up in your body over time and becomes toxic in large quantities. Put on only as needed (when heading outdoors).

Also, I would like to mention that in my analysis of popular sunscreen brands, I noticed that some of the well-advertised brands *cough cough* Banana Boat and Hawaiian Tropic *cough cough* had sunscreen products that contained parabens. If you are not familiar, parabens have been documented as potential carcinogens, as well as reproductive disruptors (Darbre et al., 2004 | Tavares et al., 2009). Something else to bear in mind when reading the sunscreen labels.

Now, this blog post is not meant to scare you and turn you into cave people that never leave their homes. But rather to provide some insight into the congested sunscreen aisle. So what is recommended for sun safety? Because lack of sun protection and melanoma are still very present dangers. Some scientists debate that some of the active ingredients are not truly harmful, as there is limited evidence and it is not a proven widespread issue ("Can the chemicals in sunscreen cause cancer?" n.d.). So in a pinch, you can use up the sunscreen you have in the cupboard (so long as it isn't expired). As the benefits of these sunscreens still outweigh the potential risks ("Debunking Melanoma Myths," 2016). This is just one field that will require much more research as we move forward. But, if you are looking for a sunscreen that is safe for you as well as safe for the coral reefs (while you are snorkelling on vacation) and the environment, look for one that uses the active ingredients zinc oxide and titanium dioxide (Morison & Wang, n.d. | Hart, 2013). So everyone, stay sun safe this Summer and remember to Keep Looking for the Science in Things.

What the SPF?

(Originally Published June 19, 2018)

So the first day of Summer for all of us here in the Northern Hemisphere is this Thursday, June 21st. And with Summer here, we are slathering up the sunscreen. If you haven't taken a look at the ingredients list on your sunscreen, you should start with our post, Can Sunscreen Really Give You Cancer? I'm sure you have seen on the bottle an SPF rating, whether it is SPF 15, 30, 50, or higher. But what does SPF mean? Is it how many minutes the sunscreen is effective for? Until recently, I thought the higher the SPF, the better the protection. But protection from what and how? Today, we are going to discuss what SPF is and how to choose the best SPF for you.

First, it's important to mention why we are wearing sunscreen in the first place. The Sun produces ultraviolet radiation (or UV rays), which go through the Earth's atmosphere, hitting everything on the surface, us included. However, we have to be careful of our exposure to these rays, in particular UVA and UVB. These types of UV radiation cause everything from sunburns (UVB) to premature skin aging and wrinkles (UVA) to skin cancer (UVB). That is why it is important to use a sunscreen that has broad-spectrum protection.

So, what's the deal with SPF? A recent interview with Dr. Steven Q. Wang, M.D., chair of the Skin Cancer Foundation Photobiology Committee, finally gave us some clarity ("Ask the Expert," 2018). Apparently, the SPF number, be it 15, 30, or 50, refers to the amount of

UVB protection that that sunscreen provides. Rather than comparing the SPFs to each other, like we all do at the store, SPF is a reflection of the length of time it would take for the Sun's UVB radiation to redden your skin (used exactly as directed) versus if you didn't apply any sunscreen at all. In ideal situations (in lab settings), if you wore SPF 30, it would take 30 times longer for you to get a sunburn than if you didn't wear any sunscreen.

What's more, SPF 30 is not nearly half the strength of SPF 50. Rather, SPF 30 allows 3% of UVB rays to hit your skin, and SPF 50 allows about 2% of UVB rays to hit your skin. Now before you say that that is just one measly percent, it actually is much more. According to Dr. Steven Q. Wang, SPF 30 allows around 1.5 times more UV radiation onto your skin than SPF 50 ("Ask the Expert," 2018). That's an actual 150% difference in protection.

Now, just because you have decided to increase the SPF protection on your sunscreen does not mean that you can go longer between applications. That's where most people end up getting sunburned more often. No matter what SPF you use, you should be applying 2 tablespoons of sunscreen 30 minutes before you head outside and be reapplying every 2 hours or immediately after swimming or sweating (even if your sunscreen is waterproof) ("Ask the Expert," 2018). Also, there are some instances where you may want to look at SPFs higher than 50. For those who have a history of or are at a higher risk for skin cancer, have skin disorders/ailments (such as albinism and skin hypersensitivity), or have immune disorders, SPF 50 may not provide enough protection. Also, if you are doing activities at higher altitudes, such as skiing and hiking, or vacationing near the Equator, you may want to look at higher SPFs than you generally use at home.

Still concerned about your skin? Well, there are other ways that you can stay sun-safe this Summer. These include limiting our exposure to UV rays by staying out of the Sun or indoors when it is at its highest (10:00 to 14:00), wearing hats and protective clothing, and seeking out shade. So as you are out and about this Summer, whether you are at the beach, hiking in the forest, working in your garden, exploring your city, or far away, remember to look after yourself and your skin. Thanks for tuning into today's post, on our new posting schedule. And remember, Keep Looking for the Science in Things.

Vegan vs. Meatarian: The Science of Food

(Originally Published July 5, 2018)

Health vegans have always perplexed me. I understand the principle of not consuming anything with a face. It's a moral thing, like I'm against animal testing and the unnecessary use of animal products in our household goods, like shampoo, lotion, soaps, cosmetics, etc. I can even understand people's objections to the practices at big commercial dairy farms. But that's a matter of ethics and procedures. Growing up on a farm with free-range chickens that laid unfertilized eggs (roosters were separated) when they felt like it (and no, we didn't kill the chicken because it stopped laying eggs) and friends who have a family dairy cow that comes in from the field to be milked, I didn't understand the need for veganism. Rather, someone could just be more ethical about their shopping, buy only free-range eggs from sources they trust and buy dairy products from smaller farms. However, there are some out there who claim to be vegan as part of a healthier lifestyle. That red meat causes cancer, but too much of anything is bad for you (one Big Mac is not going to kill you, but a Big Mac every day for your entire life and you might have problems). There are bad hormones in dairy products (again, another reason to know your sources: the Canadian dairy industry prohibits the use of recombinant bovine somatotropin (rBST), and the European Union has banned all growth hormones in both dairy and beef cattle) ("Questions and Answers," 2012). So is veganism really any healthier for you? People, like David Wolfe and other celebrity vegans, claim so, but let's get down to the facts.

What Type of Vegan Are You?

This is a question that you should be asking yourself if you are or are considering becoming a vegan. Are fruits and veggies the order of the day, or will you be relying on pasta and bread to fill you up? Both have their drawbacks. Steve Jobs was not only known for being the man behind the Apple brand but also for his take on veganism, or fruitarianism. Jobs passed away in October of 2011, and many in the medical science field have reason to believe that Jobs' eccentric diet may have played a factor in his pancreatic cancer. According to a 2007 study, they found "evidence for greater pancreatic cancer risk with a high intake of fruit juices but not with a high intake of sodas" (Nöthlings et al., 2007). Furthermore, an additional study concluded that fructose is capable of providing "the raw material" needed by cancer cells to divide and multiply (Haupt, 2013). While filming the movie jOBS, Ashton Kutcher, the method actor cast to play Steve Jobs, was hospitalised after subjecting himself to a month of Jobs' fruitarianism diet with problems with his pancreas (Haupt, 2013).

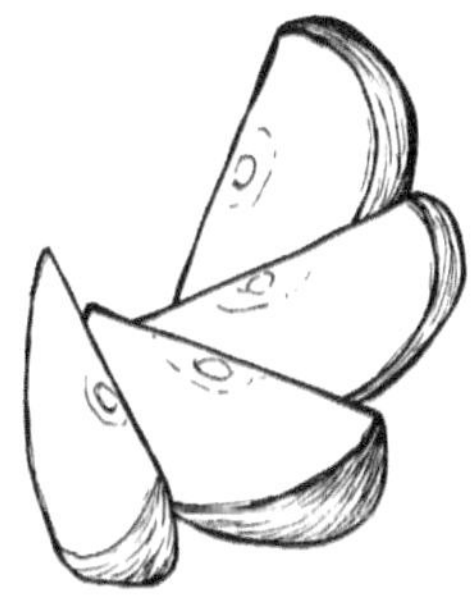

And unfortunately, a diet of bread and pasta isn't much better. Even if you eat whole wheat (I will be the first to admit I don't eat whole wheat pasta; some days I think the box that the whole wheat pasta came in would taste better) and you make it yourself, you can't escape carbohydrates. Carbs are made up of sugar, starch (another type of sugar), and fibre, which is why a meal rich in carbohydrates is perfect if you plan on having an active day ("The truth about carbs," n.d.). However, if you aren't getting your daily exercise, too many carbs can lead to too much sugar in the body ("The truth about carbs," n.d.). This can lead to weight gain, high blood sugar, and diabetes.

Junk Food Vegans & Saturday Night Slip-Ups

I guarantee we all know a junk food vegan. If you don't know who they are, they are the friend that tags along with you and the gang as you head out to your local pub for dinner. You get there, and as all of you are ordering your mouth-watering burger, your vegan friend says, "I'll just have a side order of fries." And that is their order every time you go out. Deep-fried potatoes. Fries now and again aren't bad (it's all about moderation, right), but many new vegans and those that don't know how to cook vegan resort to the quick and easy. And often that means deep-fried fast food, or when they are at home, dinner is highly processed and comes out of a box. Vegan or not, processed foods scream a few things: high in sugars, saturated fats, and chemicals. All of which have been proven, time and time again, to be contributing factors to weight gain, diabetes, heart disease, and stroke, among others.

As for the Saturday night slip ups, you are out having a good time with your friends. You decided that you've danced up an appetite, so it's off to the closest fast food joint that is still open. As you step up to the counter, your vegan friend asks you to order them chicken nuggets. A recent study conducted by Voucher Code Pros in the UK discovered that 1/3 of all vegetarians order meat when under the influence (Agency, 2015). Now, depending on how often your vegan friend goes out on the town, they might as well step back on the omnivore bus. And you know what, that's ok.

Balanced Diet

Whether you are a vegan, vegetarian, omnivore, or meatarian, eating a balanced diet is the key to being healthy. When you cut out a food group, in this case, meat, eggs, and dairy, you leave a void in some nutrients that you were getting from these sources. When people opt to go meat-free and vegan, they cut out a source of protein, iron, calcium, Omega-3, and vitamins such as B12 and B6 (Ying, 2017). These nutrients are present in other foods such as beans, legumes, tofu, kale, and spinach. But you have to make sure that you are getting them in your diet and that they are not just living on your shelf, collecting dust, or wilting in your crisper. Booking an appointment with a nutritionist may prove valuable if you are considering altering your diet.

Meat Substitutes

Now, this is a part of veganism that baffles me: meat substitutes. Maybe someone can enlighten me in the comments. I understand if someone has an allergy to meat. I know a couple of people who do. Also, if you are bitten by the lone star tick, you are at risk of developing an allergy to meat, dairy products, and products that contain alpha-gal. But if you choose to be vegan for ethical reasons, why do you want to replace bacon? I despise the practice of producing foie gras, so much so that the thought of trying to eat it or something like it makes me want to toss my cookies. I'm not going to go out and look for veggie foie gras. Anyway, back to meat substitutes. They may be more harmful to you than if you ate the meat they are substituting. And there are a few reasons for that.

First and foremost, salt. According to a recent study, it has been found that many meat substitutes can contain as much as four times more salt than their meat counterparts (Poulter, 2008). The recommended daily intake of sodium (salt) for an adult is 6 grams, with children between the ages of seven and ten allowed 5 grams, and children between four and six allowed 3 grams. Two vegetarian sausages can equal as much as 5.6 grams of salt (Poulter, 2008). And too much salt can lead to high blood pressure, heart disease, stroke, and kidney problems (Harvard, n.d.).

Next on the docket is Quorn. What in the Quorn is Quorn? Quorn is made up of a fungus that is grown in vats and is used to replace meat in some meat-substitute foods. Some doctors are sceptical of this fungus product and have questioned its impacts on the human body (Briffa, April 2018). Also, it has since been proven that the fungus in Quorn is a potential allergen, with reactions occurring on first ingestion and after building up a sensitivity to it ("Quorn," n.d.). It has been linked to two deaths (Kindy, 2015).

Finally, soy. In limited quantities, this meat substitute is perfect for those wishing to get protein with less fat. I, myself, enjoy miso soup as well as agedashi tofu every now and again. But there is something to bear in mind if you start consuming tofu en masse. Soy contains phytic acid. Phytic acid is known to hinder the absorption of various vital minerals and nutrients in your body, such as calcium, magnesium, iron, and zinc (El Tinay, 1989). Found in other grains, it appears to be very abundant in soy and other legumes. Cooking does nothing to reduce phytic acid, and the only way to reduce the amount present in soy products is by fermentation. So if you have a craving for soy, go with tempeh and miso. Also, there is some controversy around soy protein isolate (which still contains phytic acid), as the process to produce it involves being treated with various acid and base solutions. Through this, it may be contaminated with aluminium, which is currently being debated as to whether or not it is linked to nervous system degeneration and Alzheimer's disease (Briffa, May 2018). In addition, soy is abundant in hormone-like phytoestrogens. This is another mixed bag with doctors and scientists, as some believe it could protect against cancer, while others think it could have adverse effects on reproduction, increase the level of declining mental function as a person ages, and reduce thyroid function, leading to hypothyroidism (Briffa, May 2018).

Rich Man, Poor Man

But what about all the vegan celebrities that I see in the media? They look great, and they boast about their limitless energy and amazing health. Yes, but do they also credit their personal vegan chef, their nutritionist, and all the supplements that they take to fill in the gaps in their diet? There's a lot more that happens behind the curtain to make vegan celebrities appear to be shining examples of vegan health. Is it possible? Yes. Is it realistic for the average Joe? Possibly. Is it going to be as easy as the celebrities portray it? Absolutely not.

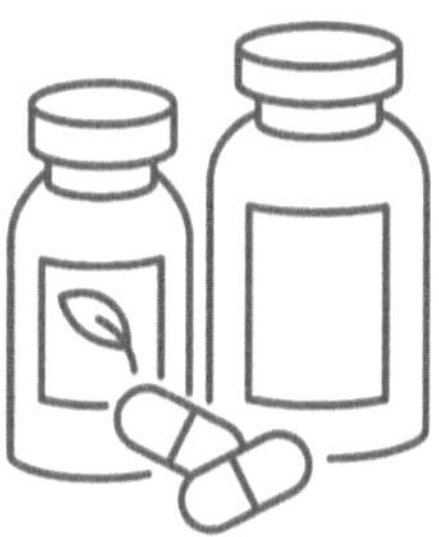

So is veganism, and by extension vegetarianism, healthier than eating meat? It can be, but it can also be worse for your health than if you eat lean, well-sourced meat. There's no harm in throwing in a Meatless Monday now and then (yes, it's a struggle in my household to have a dinner without meat present too). But whether you are vegan or not, it's essential to maintain a well-balanced diet and believe in the power of moderation to ensure you stay happy and healthy. And if you're in doubt, a trip to a nutritionist doesn't hurt either. That's all from us today. Until next time, Keep Looking for the Science in Things.

Footnote:

Since this blog post was written, a new meat alternative has appeared on grocery shelves around the world. You may recognize the names "Impossible Burger" and "Beyond Meat." But what is it exactly? These meat alternatives are plant-based and made to be comparable to their beef and chicken counterparts when it comes to protein, vitamins, and minerals (Gelsomin, 2022). The Impossible Burger gets its protein from soy, whereas Beyond Meat gets its protein from peas. Both alternatives contain vitamin B12, zinc, phytic acid, and iron, much like meat. While this may all sound great, there are some drawbacks: mainly in the form of a lot of saturated fat and sodium. Comparably, the Impossible Burger and Beyond Meat contain 8 and 5 grams of saturated fat, respectively. According to Gelsomin, 85% lean ground beef contains 6 grams of saturated fat, ground turkey contains 2 grams of saturated fat, and Sunshine Non-GMO Original brand black bean burgers contain 1 gram of saturated fat. As for sodium, the Impossible Burger has 370 mg, Beyond Meat has 390 mg, the lean ground beef has 80 mg, ground turkey has 70 mg, and the black bean burger has 300 mg. So although we have alternatives available today, it seems that we are still facing some of the pitfalls of meat alternatives that we uncovered in 2018.

Big Pharma's Big Cover Up? A Look At The Rife Machine As A Cure For Cancer

(Originally Published September 11, 2018)

Treating and finding a cure for cancer is a multibillion-dollar industry. The chances of you getting cancer in your lifetime are around 1 in 3 (depending on where you live and your lifestyle), and your chances of dying from cancer are approximately 1 in 5 (again, depending on where you live and your lifestyle) (*American Cancer Society*, 2018). Those statistics are terrifying; then you pair that with the thought of chemotherapy and radiation (the commonly prescribed means of fighting cancer); anyone who knows someone that has been through them knows the pain and illness that accompany them, and it is leading many to opt out of conventional treatment and seek alternative means. One of these alternative means that has regained some momentum is an archaic piece of machinery called the Rife machine. But what is it? And does it really work?

If you have seen the Netflix series Afflicted, you may have heard about the Rife machine, or if you follow the Gaia Facebook page, you may have seen their video. Pretty phenomenal, isn't it? Maybe a little too good to be true? Well, let's start with what we do know.

Rife and His Machine

Royal R. Rife was an American inventor (there is no evidence of him possessing a medical degree or a PhD, so we won't call him Dr.) who created a microscope in the 1930s that allowed scientists to see bacteria and viruses as they had never seen them before (*New York Times,* 1931). During his observations of microorganisms under his microscope, he developed the theory that viruses and microorganisms vibrate on their own unique frequencies, unique to their species or strain. From here, he stated that if you were able to hit that specific frequency, you could cause the virus or microorganism to burst, killing it, much like an opera singer reaching a particular pitch with their voice and shattering glass. As described in the video from Gaia, he apparently conducted tests on lab animals (sources have claimed that the tests were on lab rats) using a radio frequency emitter to attempt to rupture cancer cells and essentially kill cancer. This was apparently met with great success, so he decided to take it to the next level and begin testing on humans. This, again, was very successful, with 16 terminal cancer patients being "cured" within 90 days of the start of treatment with the newly coined "Rife machine". But rather than being a miracle cure that was celebrated around the World, Rife claims that the American Medical Association discredited him and his work and swept it under the rug. Why? Many Rife supporters believe that if cancer were cured, then the AMA and Big Pharma would no longer be able to rake in a profit to the tune of several billion dollars a year.

Big Pharma's Big Cover-Up?

Now before you get all up in arms and accuse your doctor of being a money-grubbing whatever, let's look at the facts. Gaia's video does not provide a source at the end of their video for any of the statements made, so it's time to pull up Google Scholar (the Google for all academic papers, studies, and publications for the most part). In looking there, I discovered that there are no academic papers by Rife for his work with lab rats or the clinical trials on human patients.

Down the Rabbit Hole that is Google

So with no results there, I took one giant leap down the rabbit hole and started poking around a Rife fan page known as Rife.org (not the best source for facts as fan pages can be considered biased), containing his life's work, newspaper and magazine articles, and published works, they do not present any records of published or submitted papers for the lab rat study or the human clinical trial (only frequencies for various microorganisms), which if anyone has a record of these successes (authentic or foraged), they should. You might be saying that maybe the only records were destroyed by the AMA in the 1950s in their attempt to cover up this discovery. I don't know a single scientist who does not have copies of their publications, projects, or even their lab notebooks with all their scribbles in them. I still have notes from my first year of undergrad. We are hoarders; part of recovery is admitting we have a problem, and scientists are hoarders of their work. Pair that with the fact that Rife did not pass away until 1971, and the likelihood of all copies of his work concerning the Rife machine being destroyed becomes less and less likely.

The next stop on the investigative trail is the USC website. Royal Rife supposedly conducted the human clinical trials of the Rife machine in 1934. Universities and other post-secondary institutions are incredibly proud of their famous alumni and teaching staff and actively brag about it to bring in funding, students, etc. and further build their roster (over other institutions). You can google "Princeton University" and "Albert Einstein," and you will come across multiple pages just on the Princeton website about his presence, teaching, and contributions on campus. If you go to the USC website search page, and type in "Rife," "Royal Rife," "R Rife," or "Rife, R," there is no record of him even stepping foot on campus, which is suspicious because even if he didn't invent the Rife machine, he is also supposedly noted for creating one of the most superior microscopes of the day.

Any Reliable Successes Anywhere?

Grasping at straws, I googled "Rife machine". Not recommended, but when you are looking for absolutely anything to prove that there was even a successful study done by anyone, just once, desperate times call for desperate measures. There are many papers and reviews conducted by doctors stating that the Rife machine has "big claims but no evidence." One researcher even attempted to find records of either a successful or evidence of the study of the Rife machine and its treatment of ALS, with no luck (Mac Manus, 2008 | The ALSUntangled Group, 2014). There are records of Rife machine operators being sued and documents provided by those that sell Rife machines that paint a mixed message (Barrett, 2012). One of these companies provides a document supposedly by Royal Rife describing his successes with the Rife machine, but it explains that others have been unable to duplicate his work. This leads us on a bit of a side tangent, but here goes:

> *Science Experimentation 101: In order for something to become a proven method, not just a speculation, it must have the same result time and time again, by different scientists, in different settings. E.g. If you take one raw egg and apply 200 degrees Celsius of heat to it for ten minutes, you will cook the egg, every time. You won't have a raw egg 50% of the time.*

Another interesting part of that document is that the only way you know it was written by Royal Rife is that it says "by Royal Rife." Also at the bottom, it gives you a number to call if you are interested in purchasing a Rife machine, which makes those of us that are more sceptical think that it's merely a sales tactic written by the company trying to sell you one for $1695 (especially since the Rife fan page doesn't have this document on their page).

Finally, I went to the disclaimer section of one of the Rife machine distributors, JW Labs. And if you can drudge through all the terms and agreements, you will come across this:

> *"Obviously, we cannot guarantee success or results of any kind, for any user. To do so would be contradictory to our customer agreement, and would likely be construed as fraud. We do guarantee everything we can. [...] Most failures fall into only a few categories. Lack of use being number one. Incorrect and improper use is number two. Withholding vital information is number three, which covers drug addiction, alcoholism, unhealthful habits or self destructive behavior. Other signification failures may be caused by misdiagnosis, un-treatable or incomprehensible conditions, genetic or inherited conditions, physical anomalies, and very advanced or hopeless cases that are not reported or are unidentified or unknown to the user."*

So, they can't guarantee it can or will work, and if it doesn't work, it's because of something you did or didn't do as the customer. This is a disclaimer that many therapies outside of the realm of conventional medicine must make. They cannot state that it "can cure" something or "help with" something, and in some cases, they cannot say it "alleviates symptoms". But after searching for any evidence from really anywhere, this almost seems like the final nail in the coffin in our search to prove that the Rife machine is the missing link in a cure for cancer. And until we have evidence proving otherwise, the Rife Machine probably won't cure cancer.

Now there are plenty of reviews, but how do we know if they're from someone who has used the machine or a seller? How do we know if it was the machine that cured them or something else? We don't. Big Pharma may be a business, and they may be hiding some things, but in the case of the Rife machine, there has been too much time since its invention and too little evidence of actual success (not claims) to ignore. There may be an alternative to chemotherapy and radiation out there; we just haven't found it yet. Hopefully, today's post also gave you some insight into how we find answers for you. In the meantime, Remember to Keep Looking for the Science in Things.

<u>Footnote:</u>

The disclaimer quoted from JW Labs is no longer accessible. When recently revisiting the link, for the purpose of writing this book, a 404 page in place of the former disclaimer. A new disclaimer page is on the JW Lab website which states:

> *JWLABS Machines are media adapters that act as transducers, converting audio frequencies such as the sound of music to electrical impulses, which the user feels as gentle vibrations or a tingling sensation. The JWLABS Machines are not designed to nor do we intend that they will exercise any structure-function effect on the user.*
> *These are not FDA certified medical devices and we offer them only for general well being and entertainment purposes.*

The Not-So-Secret Side of Dream Science

(Originally Published September 25, 2018)

Today is World Dream Day, and whether you dream of flying, or spending time on gorgeous white sand beaches, dreaming is a natural function of the human brain that happens every night whether you remember your dreams or not. So in today's post, we will discuss how and why we dream, as well as what happens when we can't dream.

Sleep Tight

Before we discuss dreaming, we should touch on how we sleep. A typical person sleeps in five stages, which last 5 to 15 minutes each, making a sleep cycle, which in turn lasts 90 to 110 minutes. That's why they say if you sleep in a multiple of 90 minutes (so 1.5 hours, 3 hours, 4.5 hours, 6 hours, 7.5 hours, etc.), you will wake more rested because you completed the sleep cycle, rather than being disturbed part way through. The five stages in a sleep cycle are as follows:

Stage 1: Typical in the first stage, the person is lightly sleeping, often drifting in and out of consciousness, and can be quite easily awoken. Eye movement and muscle activity slow down. However, it is not uncommon for a person's muscles to twitch, e.g. legs.

Stage 2: Eye movement stops and brain activity slows, with a very occasional, brief increase in brain waves. The body is drifting into a deep sleep at this point, with temperature lowering and heart rate slowing down to resting rate.

Stage 3: Welcome to Deep Sleep. Very slow brain waves, called delta waves, are offset by smaller, faster waves. During this stage, the individual may sleepwalk, experience night terrors, and begin talking in their sleep. Also, for parents' who have kids that wet the bed, this is typically when that occurs as well.

Stage 4: Delta waves occur more frequently in this stage, which is still a part of deep sleep. If the individual is awoken during this stage, they will often be briefly disoriented. That is why it is vital to complete a sleep cycle to feel fully rested.

Stage 5: REM (rapid eye movement) Sleep. Brain waves during this stage are very similar to those when you are awake. Even though your eyes are closed, they will move quickly from side to side; this is when dreaming mostly occurs (but not exclusively).

Sweet Dreams

Traditionally, it has been believed that we only dream during the REM stage of sleep; however, recent studies have concluded that dreaming can occur at any stage of the sleep cycle (Nir & Tononi, 2010). While you are dreaming, our entire brain is involved. Our emotions (both awake and dreaming) are processed by the limbic system, and if you are having a nightmare, you can thank your amygdala for that feeling of fear (Blackmore, 2018). As for what we dream of - the places, the people, the monsters, the cast, the setting - our cortex is the one that creates it all (Blackmore, 2018). Despite all these active parts of your brain, the frontal lobes take a backseat, which is why you are perfectly accepting of the fact that you need to go to the store for jelly beans and dish soap for your job interview (Blackmore, 2018). Your frontal lobes are responsible for critical thinking, and because they aren't as active as when you are awake, you are not analysing your dreams to see if they make sense.

Why Do We Dream?

So do we even dream? Although there is no one reason why we dream, researchers do have some theories ("Facts About Dreaming," n.d.). Some believe that dreams help us process our thoughts, emotions, and events, as well as allow us to solve problems. Others believe that dreaming helps us incorporate memories, almost like a replay of the day's events. The famous psychologist Sigmund Freud believed dreams revealed our subconscious desires, thoughts, and motivations ("Facts About Dreaming," n.d.). He also thought that dreams allowed people to fulfill wants and needs that were not socially acceptable.

Can't Dream?

We all know how important it is to sleep, and we have reason to believe that after an extended period without it, you can die. A notable study in 1989 by Allan Rechtschaffen and Bernard Bergmann, discovered that when you deprive lab rats (which have similar brain chemistry to humans) of sleep, they often die within two to four weeks unless they are allowed to sleep. So it is without question that we need our sleep, but what happens if we go extended periods of time without dreaming? Recent studies have shown that dreaming is just as important as sleeping for our physical and mental wellbeing ("Dreams and their relation to physical and mental well-being," n.d.). When someone no longer dreams, they are more open to physical and mental illness, including depression, anxiety, and other mental disorders. That is why it is essential to ensure that you take enough time to fully complete your sleep cycles, rather than being jarred awake right in the middle.

Dreaming is just as important as sleeping, despite it often being dismissed as a whimsical folly. Not only does it allow us to escape our day-to-day lives, but it also helps us to process the events of the day, how we feel, and solve that nagging problem. It has been a subject of study for psychologists, like Freud, and dreamers wanting to know the deeper meaning. Well, that's all from us for today. In the meantime, Keep Looking for the Science in Things.

Cuppa Poison?

(Originally Published November 27, 2018)

Nothing is more soothing after a long, stressful day than settling in with a nice cup of tea. However, it's important to know where your tea is coming from. But is it really that important? Isn't all tea the same? Not so, according to recent studies conducted by CBC News, Greenpeace, and Glaucus Research (Griffith-Greene, 2014 | "TROUBLE BREWING," 2014 | *Glaucus Research Group*, 2013). Your favourite cup of tea might contain pesticides as well as a slew of other chemicals. Have no fear; we got the tea scoop, so you can be better informed next time you head to the grocery store.

What's In Your Cup?

During the CBC News study, they analyzed various popular brands and types of tea, testing for pesticide residue on the dried tea levels at an accredited lab, following methods that are used by the National Food Inspection Agency. And the results were startling! The majority of the teas tested had pesticide residue above the legal limit. Numerous chemicals were found in most of the teas tested, and one brand had a startling 22 different types of pesticides present in their tea (Uncle Lee's Legends of China). Most of the pesticides found in the tea are in fact banned in multiple countries based on the negative impacts they have on the environment, as well as the health impacts on the workers that handle them. And we are drinking them. Yikes!

In addition to CBC and Greenpeace's studies, Glaucus Research discovered that around 91% of Celestial Seasonings tea contained pesticide residue that exceeded the U.S. legal limit. Many of which contain known carcinogens.

Chemical Culprits

Here's the list of teas that tested positive for pesticides and various chemicals. So if they are in your cupboard, throw them out! Even my cupboard is getting cleaned out.

Brooke Bond
- Red Label
- Red Label Natural Care
- Red Label Special
- 3 Roses Natural Care
- Taj Mahal

Celestial Seasonings
- Authentic Green Tea
- Antioxidant Max Blackberry Pomegranate
- Antioxidant Max Blood Orange
- Antioxidant Max Dragon Fruit
- English Breakfast Black K-Cup
- Green Tea Honey Lemon Ginger
- Green Tea Peach Blossom
- Green Tea Raspberry Gardens
- Sleepytime Herbal Teas (Flagship)
- Sleepytime Kids Goodnight Grape Herbal

Golden Tips
- Nilgiri Tea
- Pure Darjeeling Tea
- Assam Tea

Goodricke
- Chai Strong CTC Long Leaf
- Roasted Darjeeling – Orange Pekoe
- Thurbo Flavoury Darjeeling Tea
- Thurbo Flavoury Darjeeling Tea

Kanan Devan

Kho Cha
- Darjeeling
- Masala Chai

King Cole
- Orange Pekoe

Lipton
- Clear Green Tea
- Darjeeling Tea
- Pure Green Tea
- Yellow Label Black Tea

No Name
- Black Tea

Royal Girnar Cup Tea Signal
- Orange Pekoe Two Cups

Tata Tea
- Tata Tea Gold
- Tata Tea Life
- Tata Tea Premium

Tetley
- Pure Green Tea
- Long Leaf Green Tea

Twinings
- Classic Assam Tea

    – Classic Lady Grey
    – English Breakfast
    – Earl Grey

Uncle Lee's Legends of China
    – Green Tea
    – Jasmine Green Tea

Wagh Bakri
    – Good Morning Tea
    – Perfect Premium Leaf Tea
    – Strong & Refreshing Premium Leaf Tea

I personally pitched a few more Tetley and Twinings teas in the garbage, even those that weren't on the list, because, as the saying goes, When in Doubt, Chuck It Out. (Yes, that is a mantra about leftovers in the fridge, but I think it fits here.)

Breakdown of the Pesticides

So what cocktail of pesticides were found in these teas, and what are their impacts on humans? We've got the full list from the CBC study right here:

One thing to note with the description of these pesticides is the hierarchy of toxicity:

Fatal > Toxic > Harmful

You Are Dead > This Is Really Gonna Suck > You're Not Gonna Like This

Acephate – An organophosphate insecticide that kills target insects when they touch or eat it (Christiansen et al., 2011). When sprayed on plants, the plant absorbs acephate into its system. According to the EPA, acephate is classified as a "possible human carcinogen," as multiple repeat exposures and ingestions have resulted in liver or adrenal gland tumours in test mice, as well as DNA damage in the blood cells in mice that were fed high doses of acephate. In addition to these effects, the test rats also experienced reduced fertility and increased offspring mortality.

Acephate

Acetamiprid – A neonicotinoid insecticide, that kills insects by acting as a neurotoxin that overwhelms their nicotinic receptors (National Center for Biotechnology Information, "Acetamiprid," 2011). Although it is classified as an unlikely human carcinogen, it should be noted that it is toxic if ingested, and some nervous system-related symptoms (e.g. seizures) have occurred in humans. In lab rats, there were also cases of reduced fertility and increased mortality in offspring, as well as lower body weights in individuals that ingested acetamiprid.

Acetamiprid

Beta-Endosulfan – An organochloride insecticide that kills insects that suck, chew, or bore into plants (University of Hertfordshire, 2018). It can be absorbed through ingestion, inhalation, and skin contact. Although it is an unlikely carcinogen, it is known to be a possible endocrine disrupter and neurotoxicant, as well as having potential negative reproductive and developmental issues.

Beta-Endosulfan

Bifenthrin – A broad-spectrum insecticide (National Center for Biotechnology Information, "Bifenthrin," n.d.). Bifenthrin can be absorbed into the human system through ingestion, inhalation, or the skin. Classified as a poison, it is fatal if ingested, can cause skin irritation, is toxic if inhaled, is a possible carcinogen, and is known to cause organ damage due to prolonged and/or repeated exposure. Not only does it have negative human effects when ingested, inhaled, or comes into contact with skin, but when heated up, it also emits the toxic gases chlorine and fluoride.

Bifenthrin

Buprofezin – An insecticide that kills sap-sucking insects by inhibiting their ability to produce chitin, which makes up their exoskeleton (National Center for Biotechnology Information, "Buprofezin," n.d.). It is not a known carcinogen, however, it can cause organ damage when exposed to it for a prolonged period of time and/or repeatedly. Also, when heated up, it emits toxic gases like nitrogen oxide and sulfur oxide.

Buprofezin

Carbendazim – A broad-spectrum fungicide (National Center for Biotechnology Information, "Carbendazim," n.d.). As a possible carcinogen, it has also been known to cause genetic defects, reduce an individual's fertility and harm unborn children. When heated up, it emits nitric oxide, which is toxic.

Carbendazim

Chlorfenapyr – An organochloride and organofluoride insecticide (National Center for Biotechnology Information, "Chlorfenapyr," n.d.). There has been some evidence of carcinogenic properties, but not enough to definitively conclude the possibility of cancer in humans. Despite this, it is harmful if swallowed, can cause eye irritation, is toxic if inhaled, may cause organ damage, and causes organ damage as a result of prolonged and/or repeated exposure. When heated up, it emits hydrogen chloride, hydrogen fluoride, and nitrogen oxide, all of which are toxic.

Chlorfenapyr

Chlorpyrifos – An organothiophosphate insecticide (National Center for Biotechnology Information, "Chlorpyrifos," n.d.). It is one of two higher-risk insecticides for lung cancer. Long-term exposure has led to the presence of autoimmune antibodies in those with autoimmune disorders. It is toxic if swallowed, harmful if it comes into contact with skin, can cause serious eye irritation, and is fatal if inhaled.

Chlorpyrifos

Clofentezine – An organochloride acaricide that is commonly used to kill mites (National Center for Biotechnology

Information, "Clofentezine," n.d.). It is a known carcinogen and is harmful when it comes into contact with skin.

Clofentezine

Cypermethrin – A pyrethroid ester insecticide, acaricide, and molluscicide (National Center for Biotechnology Information, "Cypermethrin," n.d.). It is not a known carcinogen; however, it is toxic if swallowed, harmful if inhaled, known to cause respiratory irritation, and causes organ damage when exposed for long periods of time and/or repeatedly.

Cypermethrin

Deltamethrin – A pyrethroid ester insecticide most commonly used in killing malarial mosquitoes (National Center for Biotechnology Information, "Deltamethrin," n.d.). It is the most popular and widely used insecticide in the World. And yet, it isn't without its risks. It is toxic if swallowed, can cause allergic skin reactions and serious eye irritation, is toxic if inhaled, may cause respiratory irritation, is suspected of decreasing one's fertility as well as harming unborn children, and causes organ damage when exposed for long periods of time and/or repeatedly. In addition, it can produce the toxic fumes hydrogen cyanide and hydrogen bromide when heated to high temperatures.

Deltamethrin

Dicofol – An organochlorine insecticide (National Center for Biotechnology Information, "Dicofol," n.d.). You may better know the chemical concoction from which it is derived, DDT. It is harmful if swallowed and capable of causing allergic skin reactions, serious eye damage, and organ damage, including organ damage caused by long and repeated exposures.

Dicofol

Dimethoate – An organophosphate insecticide that has a wide array of uses as an insecticide because of its potent effects on the central nervous system (National Center for Biotechnology Information, "Dimethoate," n.d.). It is toxic if swallowed or in contact with skin, causes eye irritation, and causes organ damage, including respiratory failure. It is also known to negatively affect the development of embryos.

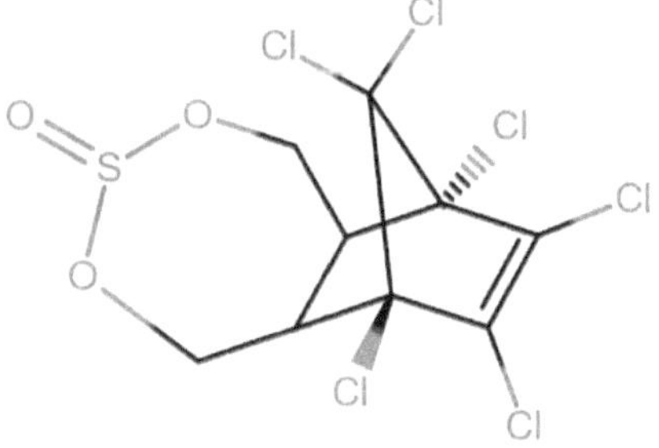

Dimethoate

Endosulfan – A pesticide that is used on insects, fungi, and plants (National Center for Biotechnology Information, "Dimethoate," n.d.). It is a very persistent pesticide that does not dissolve in water and sticks to clay and soil particles. It is toxic if swallowed, inhaled, or comes into contact with skin.

Endosulfan

Etofenprox – A pyrethroid ether insecticide (National Center for Biotechnology Information, "Etofenprox," n.d.). It could cause harm to breast-fed children and cause organ damage when exposed for long periods of time and/or repeatedly.

Etofenprox

Fenazaquin – An acaricide used to kill mites and ticks (National Center for Biotechnology Information, "Fenazaquin," n.d.). It is toxic if swallowed and harmful if inhaled.

Fenazaquin

Fenpropathrin – A pyrethroid ester insecticide and acaricide (National Center for Biotechnology Information, "Fenpropathrin," n.d.). It is toxic if swallowed, harmful if it comes into contact with skin, and fatal if inhaled.

Fenpropathrin

Fenpyroximate – A pyrazole acaricide (National Center for Biotechnology Information, "Fenpyroximate," n.d.). It is toxic if swallowed, fatal if inhaled, can cause allergic skin reactions, and causes organ damage from long periods of exposure or repeated exposure.

Fenpyroximate

Fipronil – An insecticide that is also used in veterinary medicine to kill parasites (National Center for Biotechnology Information, "Fipronil," n.d.). However, for people, it is toxic if swallowed, toxic if it comes into contact with skin, can be fatal if inhaled, and causes organ damage when exposed for long periods of time or repeatedly.

Fipronil

gamma-Cyhalothrin/lambda-Cyhalothrin – A pyrethroid ester insecticide (National Center for Biotechnology Information, "gamma-Cyhalothrin," n.d.). Although not a known carcinogen, it is toxic if swallowed, toxic and harmful if it comes into contact with skin, causes serious eye irritation, is fatal if inhaled, and can cause organ damage. In addition to these concerns, it also produces nitrogen oxides, hydrogen chloride, and hydrogen fluoride gases when burned.

gamma-Cyhalothrin

lambda-Cyhalothrin

Hexythiazox – An insecticide and acaricide (National Center for Biotechnology Information, "Hexythiazox," n.d.). It has no reported adverse effects on humans; only slight eye irritation was reported in lab tests. However, it is very toxic to aquatic life.

Hexythiazox

Imidacloprid – A neonicotinoid insecticide (National Center for Biotechnology Information, "N-[1-[(6-chloro-3-pyridinyl)methyl]-4,5-dihydroimidazol-2-yl]nitramide," n.d.). It is harmful if swallowed (no other adverse effects on humans have been recorded). However, it is very toxic to aquatic life.

Imidacloprid

Methomyl – A carbamate insecticide and nematocide (kills worms) (National Center for Biotechnology Information, "Lannate," n.d.). It is fatal if swallowed or inhaled and is also known to cause drowsiness and dizziness. When heated, it releases the toxic gases nitrogen oxide and sulfur oxide.

Methomyl

Monocrotophos – An organophosphate insecticide (National Center for Biotechnology Information, "Monocrotophos," n.d.). It is very genotoxic, meaning that it has a very destructive effect on genetic material, including genes and DNA. Also, it is fatal if swallowed or inhaled, and it is toxic if it comes into contact with skin.

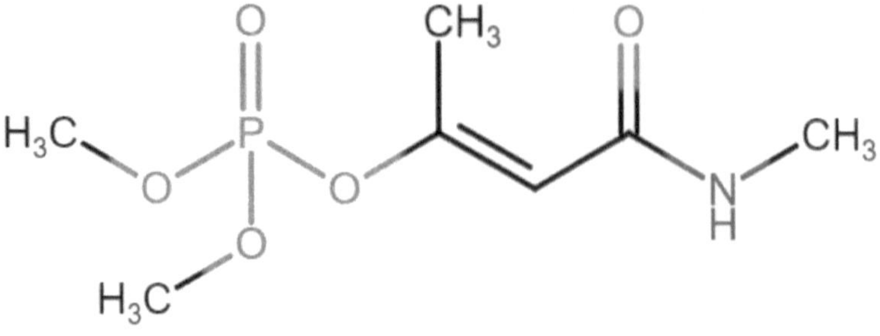

Monocrotophos

Omethoate – An organothiophosphate insecticide, quite similar to dimethoate (National Center for Biotechnology Information, "Omethoate," n.d.). It is fatal if swallowed and toxic if it comes into contact with skin.

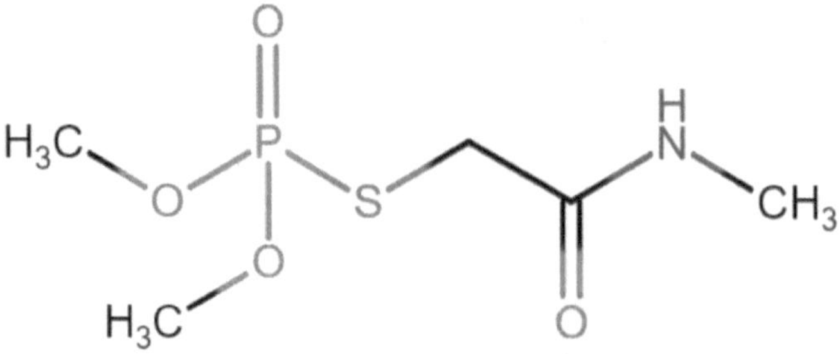

Omethoate

Propargite – An insecticide and acaricide (National Center for Biotechnology Information, "Propargite," n.d.). A suspected carcinogen. It is toxic if inhaled and causes skin irritation and serious eye damage.

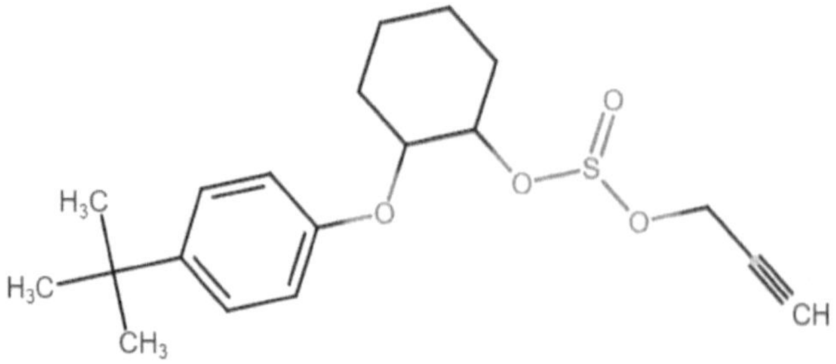

Propargite

Pyridaben – An organochlorine insecticide and acaricide (National Center for Biotechnology Information, "Pyridaben," n.d.). It is toxic if swallowed or inhaled.

Pyridaben

Thiacloprid – A neonicotinoid insecticide, as well as a known environmental contaminant (National Center for Biotechnology Information, "Thiacloprid," n.d.). A suspected carcinogen. Toxic if swallowed and harmful if inhaled, it may cause drowsiness and dizziness, reduce one's fertility, and harm unborn children.

Thiacloprid

Thiamethoxam – An insecticide and fungicide (National Center for Biotechnology Information, "Actara," n.d.). It is harmful if swallowed or inhaled. It is also a highly flammable solid.

Thiamethoxam

Thiodicarb – An insecticide (National Center for Biotechnology Information, "Thiodicarb," n.d.). It is toxic if swallowed, causes eye irritation, is suspected to decrease one's fertility and harm unborn children, and causes organ damage, including through prolonged and repeated exposure.

Thiodicarb

Triazophos – A pesticide used to control insects, mites, and nematodes (National Center for Biotechnology Information, "Triazophos," n.d.). Despite this use, it is not registered as a pesticide in the US. It is toxic if swallowed or inhaled, and it is harmful if it comes into contact with skin.

Triazophos

This list only includes the effects these chemicals have on humans. Many of these chemicals also have devastating consequences for the environment.

Out of all these chemicals found in tea, only bifenthrin, dicofol, fenazaquin, fenpropathrin, fenpyroximate, propargite, and thiamethoxam are approved for use on tea plants ("TROUBLE BREWING," 2014).

One thing to note about pesticides is that children are much more sensitive to them, and they are experiencing more symptoms, with greater severity.

Other Toxic Chemicals Found in Your Cup

In addition to all the pesticides and chemicals found on the dry tea leaves, we may be getting poisoned by the tea bags themselves. How often have you ripped open a sachet of tea and pulled out a stark white tea bag? Yup, the tea bag is bleached. In addition to the bleach, some of the paper tea bags contain epichlorohydrin, a pesticide that prevents the tea bags from dissolving in hot water (Fraser, 2016).

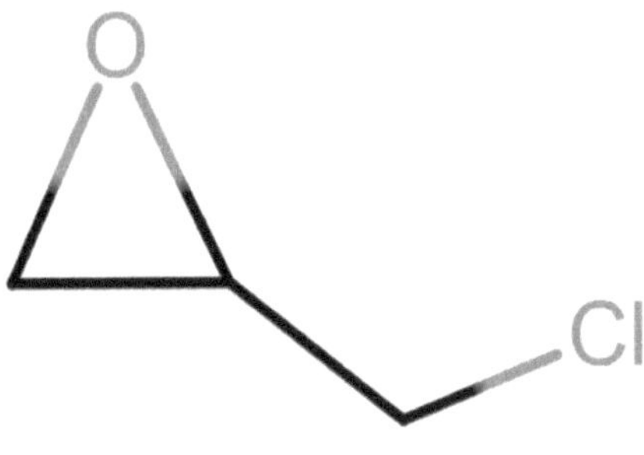

Epichlorohydrin

Think those mesh tea bags are any better? Think again. They are made from a corn-based plastic known as polylactic acid, which, when placed in boiling water, leaches harmful chemicals, including BPA, into your tea (Fraser, 2016).

Be An Informed Shopper

Just because of these latest discoveries, it doesn't mean that you have to give up tea altogether. You just need to be an informed shopper. Here are some tips to avoid drinking a toxic cocktail rather than a soothing cup of tea:

1. Buy organic-certified brands of tea.

2. Know your tea sources. Where do the ingredients come from, and what are those companies' or country of origin's practices?

3. Check your ingredients. There should be nothing in them that you cannot read. No chemicals. No artificial flavours. No weird preservatives.

4. Purchase loose-leaf tea. No tea bag = no bleach, no epichlorohydrin, no polylactic acid.

5. If you use tea bags, look for ones that are unbleached, organic, and do not contain chemicals.

In a world full of toxic chemicals that are often used inappropriately, it has become our responsibility to be informed consumers. To know what is in the items that we buy, the effects they have on us, our families, our home, and our environment. That's all from us today. In the meantime, Keep Looking For the Science in Things.

<u>Footnote:</u>

While compiling this book, I decided to revisit the brands of tea that wound up on our radar due to their high levels of pesticides that had been discovered at the time of writing this blog post. Here's how they fair now:

- **Brooke Bond** - According to RateTea.com, the Brooke Bond brand is currently owned by Unilever and has been mostly discontinued in the Western Market, except for the occasional South Asian import store. There have been no reports on whether Brooke Bond has changed their pesticide use practices; however, following GreenPeace's report, Hindustan Unilever Ltd.

announced a statement of their commitment to a gradual phase-out of pesticides, according to GreenPeace.

- **Celestial Seasonings** - Upon release of the initial results, Celestial Seasonings supposedly sent their own samples to the National Food Laboratory and released their own statement. However, when trying to find their statement at the time of writing this footnote, all links to said statement just lead to a 404 error page. Supposedly, their statement included the following: "NFL's independent testing reaffirmed that Celestial Seasonings teas are safe and follow strict industry guidelines. In addition, the NFL detected no pesticides in the brewed Celestial Seasonings teas they tested." However, I did manage to find a statement on their website stating: "All incoming raw ingredients are tested for purity and quality. Any raw ingredients that do not meet the standards established by leading government and industry agencies, including the U.S. Food and Drug Administration and the European Union Pharmacopoeia Convention, are rejected." But without access to their test results, I can neither confirm nor deny the presence of pesticides in Celestial Seasonings' tea at this time.

- **Golden Tips** - No statements or press releases have been released by Golden Tips regarding pesticides at the time of writing this footnote.

- **Goodricke** - On a cursory search of Goodricke's website, I was able to find a statement regarding their continued efforts to reduce their pesticide load, and they only use approved and safe pesticides. However, they failed to state what pesticides they do use or provide any lab results for their products. No statements have been made by Goodricke about the pesticide allegations.

- **Kanan Devan** - similar to Goodricke, Kanan Devan's website provides a statement that they provide safe and quality products, but there are no attached lab reports. No statements have been made by Kanan Devan about the pesticide allegations.

- **Kho Cha** - I was unable to find a website or social media accounts for Kho Cha. No statements have been made by Kho Cha about the pesticide allegations.

- **King Cole** - King Cole's website provides no statement to the alleged report of pesticides in their tea, nor do they mention any quality assurance or methods for assuring that their product is safe other than having a tea taster named Des McCarthy.

- **Lipton** - On Lipton's website, under the FAQ section, they actually address the concerns about the presence of pesticides in their tea. "We have committed that by 2020 all our food raw materials will be produced using sustainable crop practices, minimizing the use of pesticides through integrated pest management techniques, and with due care for the environment and the livelihood of farmers... Minimizing use of pesticides is always a requirement. We monitor the presence of pesticide residues in the raw materials that we purchase and we take action with our suppliers to ensure that the ingredients continue to meet the regulations in force. We are constantly reviewing our pesticide approach and practices in light of new scientific evidence and societal concerns here." However, no lab reports have been included with their statement at this time.

- **No Name** – Neither No Name nor their parent company, Loblaws, have issued a statement regarding the alleged presence of pesticides in their products. On No Name's website, there is no specific statement about the quality of their tea or their methods for ensuring their products safety.

- **Royal Girnar Cup Tea Signal / Girnar** - Following the release of GreenPeace's report, Girnar released a statement of their commitment to reduce their pesticide usage, according to GreenPeace. However, this commitment is not listed on their website. They do, however, mention that they have a "No Human Touch" process, meaning that only machines carry out the blending, packing, and sealing procedures.

- **Tata Tea & Tetley** - This one I am going to chalk it up to something new I learned today: Tetley and Tata Tea are owned by the same parent company, Tata Consumer Products, which is in turn owned by the Tata Group. Although the Tata Consumer Group did not specifically address the alleged pesticides in their products, they do make a statement on their website about their efforts to reduce their dependency on chemicals and minimise the negative effects on the environment. They go one step further and mention the efforts that their suppliers make to reduce their use of harmful chemicals. However, no lab reports have been provided on their website to back up their claims at this time.

- **Twinings** - I hate to say it, but on a quick search of Google and Twining's website, I couldn't find any statements regarding their stance on pesticide usage. Which is unfortunate because of all the companies that I looked at for the footnotes, they have the most information about their initiatives to provide better conditions for their suppliers and reduce their impact on the environment (zero carbon footprint).

- **Uncle Lee's** - Uncle Lee's tea has been certified organic by the USDA and QAI, both of which are independent bodies. According to the USDA, in order for a product to be considered organic, it must be free "of most synthetic pesticides and fertilizers, growth hormones, sewage sludge, irradiation, and genetic engineering (genetically modified organisms or GMOs)." So that seems promising. In addition, I was able to pull the organic certification records from QAI. The most recent certification for the Green Tea was February 15, 2012 (which was prior to the CBC, GreenPeace, and Glaucus investigations), and for the Jasmine Green tea, it was June 3, 2019. And that's where Uncle Lee's falls off the rails for me. Without a more recent test of their green tea, I cannot confirm if it is still organic and free of chemical pesticides.

- **Wagh Bakri** - Following the release of GreenPeace's report, Wagh Bakri responded with the following: "Wagh Bakri has committed to invest in pilots, along with other stakeholders to facilitate the development and evaluation of non chemical crop protection management with an approach of holistic rejuvenation of the ecosystem." This led me to look at their company website, which has a page of their certifications, which are almost impossible to read. From here, I discovered that their company also operates under the name of Gujarat Tea Processors and Packers Ltd. When looking for confirmation of their USDA NOP certification, I could not find any records under the names of Wagh Bakri or GTPPL. However, when looking up their BRCGS certification, I found that they had received a Food Standard grade of A from Intertek Certification Ltd., with a valid certificate from April 18, 2023, to May 14, 2024. For those of you wondering what this certification means, it involves managing "product safety, integrity, legality and quality, and the operational controls for these criteria in the food and food ingredient manufacturing, processing and packing." It does not specifically pertain to pesticides, but it is a definite start.

Math + Physics

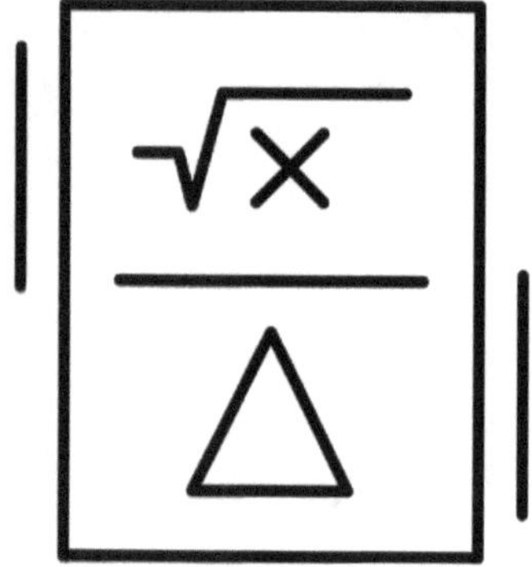

Paper Plane Physics

(Originally Published May 21, 2018)

I know pretty much everyone, as children, used to make paper airplanes at some point in time. And whether they were those long-distance flyers or whether they did loop-de-loops, I would always go back to tweak it, whether it was to stop it from flying loop-de-loops, fly bigger loops, or fly further. Little did I know, I was actually experimenting with physics and engineering. So in today's post, we are going to talk about paper airplanes, what makes them fly great distances, and why some do loop-de-loops.

Paper airplanes are a basic representation of a glider, as they don't have an engine or a source of thrust to maintain flight after the initial launch ("Paper Airplanes," 2015). Like a glider, paper airplanes have three forces acting on them at any given time (power aircraft have four). These forces are drag, lift, and gravity. All these forces are created as a result of a solid object moving through a fluid or fluid-like substance. For example, a boat moving through water or a plane in the air.

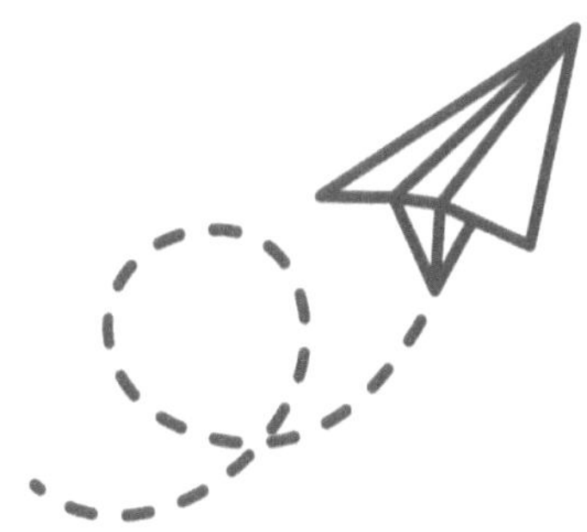

Drag

Drag acts in the opposite direction to which an object is going and is often responsible for an object slowing down when it doesn't have the energy to continue, much like friction ("What is Drag?" 2015). In paper airplanes, it is responsible for why the plane slows down and is partially responsible for it hitting the ground. The amount of drag that an object experiences is dependent on its shape and, therefore, how aerodynamic an object is. An upright billboard will experience a lot more drag than a bullet. With paper airplanes, you want to minimize your drag as much as possible if you wish to achieve faster speeds and further distances. Too much drag, and it won't go very far.

Lift

Lift acts perpendicular to the air flow or direction in which the object is going ("What is Lift?" 2015). It is created by a difference in air pressure above and below the object. Birds' and airplanes' wings have a different shape on top versus underneath in order to maximize their lift capabilities. The top of the wing is shaped to be as aerodynamic (and minimize drag) as possible, creating a lower air pressure above the wing; whereas, the underside of the wing is shaped to slow down the air, creating a higher air pressure on the wings. With this difference in air pressure, objects as heavy as 747s can get off the ground. It is also why, no matter how hard we flapped our arms as kids, we would stay firmly on the ground. When your paper airplane has too much lift, it will perform loop-de-loops as it keeps a continuous lower air pressure above the wing than below it. As you may have noticed, airplanes have wing flaps to control the amount of lift that they have, generally extending and adjusting their angle during liftoff and landing. If there is not enough lift, and it nosedives straight after you throw it.

Gravity

Gravity is the final force that acts on gliders and paper airplanes, as it does on pretty much everything on Earth, including the Earth itself. Gravity is a force that is dependent on an object's weight and acts in the opposite direction of lift ("What is Weight?" 2015). So the heavier something is, the more gravity is acting on it. So next time you step on the scale and aren't happy with the numbers you see, don't get mad at the scale and think it's broken; blame gravity. Also, the more gravity something has acting on it, the more it will want to return to the ground. That's why when you drop a bowling ball from chest height, it will crash to the floor, and a foam ball dropped from chest height will take a fraction of a second longer (terminal velocity puts a kink in this, but we will discuss that another day). When looking at your paper airplane, you are going to want to keep it lighter the smaller the surface area of your wings are. Too heavy, and it will go just as far as if you crumpled it into a ball and threw it. Too light, and you may again have a plane that does loop-de-loops.

Thrust

A final force that is worth mentioning is thrust ("What is Thrust? 2015). Thrust is the force that is propelling the object forward. In airplanes, it is their engines. Paper airplanes have an initial thrust moment, which is when you are throwing them out of your hand. The obvious difference in thrust between a paper airplane and an actual airplane is that the paper airplane, after being thrown, does not continue to have thrust.

When you get the right balance of all the forces acting on your paper airplane, you will get a long-distance flyer. Just like Ken Blackburn's World record paper airplane flight of 27.6 seconds. That's all for now. Keep Looking for the Science in Things.

Luck of the Draw

(Originally Published June 11, 2018)

Do you know someone who seems to just have all the good luck? Whether it is landing the dream job or picking the winning lottery numbers, it just seems like no matter what they do, everything works out in their favour. It may seem like they have an unlimited number of horseshoes shoved up their rear end, have a lucky rabbit's foot, or have a four-leafed clover in their wallet. But what if I told you that there is no such thing as luck, but rather just a mathematical or statistical concept called probability and chance.

There have been a few movies lately that have highlighted luck. Deadpool 2 features Domino, a heroine who claims her superpower is luck. Despite Deadpool's claim that luck is not a superpower, Domino has an uncanny ability to make everything work out in her favour. So what is luck, anyway? Well, luck is actually dependent on probability and chance, and it doesn't become lucky or unlucky until someone applies it personally. For example, there is a chance a vehicle will be stolen if it is left unlocked. It is considered unlucky if your vehicle is stolen, or lucky if your vehicle is still there and your wallet is still left on the driver's seat. Luck (and unluck) is a concept that we, as humans, have come up with to help describe an event's likelihood and is more psychological than actual fact (Ossola, 2015). So let's take a look at some events that are deemed lucky or unlucky and see what the probability pof them actually happening is.

Coin Flip

We'll start off with a basic one that is commonly studied: the coin flip. Now we all know that the chance of flipping a coin and getting "heads" is 50/50 (Ossola, 2015). And each event is independent of one another, so if you were to flip a coin a second time and get a "heads," it is 50/50. So the chance of getting two "heads" in a row would be 50/50 again, right? Wrong. If you are looking at the chance of getting two "heads" in a row, you then have to look at the events as a series:

Chance of getting heads #1 x Chance of getting heads #2 = Chance of getting heads twice in a row

0.5 x 0.5 = 0.25 or 25% chance

As you can see, the chance of getting two "heads" in a row is 25% rather than 50/50 (or 50%). And you would see that as you apply more conditions (eg. chance of getting heads a third or fourth time in a row), the likelihood of that event occurring would decrease as well.

Rolling Snake Eyes

Common in board games and gambling, rolling snake eyes, for anyone that is not familiar with the term, is when two six-sided dice are rolled and they both show 1s. This is another "lucky" event that we can also calculate:

Chance of getting 1 x Chance of getting 1
= Chance of getting snake eyes

1/6 x 1/6 = 0.17 x 0.17 = 0.03 or 3%

So the next time someone rolls a set of snake eyes in Monopoly, know that they only had a 3% chance of rolling that end result.

Lottery

Millions of people around the World purchase lottery tickets every day, hoping to get lucky and win millions. But we all now know that luck has nothing to do with it, and rather, these individuals are betting their odds at the chance to strike it rich. In Canada, there are multiple lotteries drawn every week, but one of the most notable is the Lotto Max, which draws every Friday. This past Friday (June 8, 2018), someone walked away with $60 million. What were the odds of them being one of the luckiest people that day, versus every other unlucky person that did win? According to the Western Canada Lottery Corporation's website, Lotto Max is a lottery draw of seven numbers ranging from 1 to 49. Everyone who plays picks (or has the numbers randomly assigned to them) seven numbers. So, with that in mind, let's do the math:

Number 1 x Number 2 x Number 3 x Number 4 x Number 5 x Number 6 x Number 7
= Chance of having the winning number combination

1/49 x 1/48 x 1/47 x 1/46 x 1/45 x 1/44 x 1/43
= Chance of Winning

0.0204 x 0.0208 x 0.0213 x 0.0217 x 0.0222 x 0.0227 x 0.0233
= 0.00000000002302860297282854 or
000.00000000023028602972828544% chance of winning

In case you didn't notice, the odds of your number being drawn actually go up as each number is drawn. This is because if a number is drawn, it is not re-entered back into the pool of numbers; therefore, there are fewer and fewer numbers to choose from. So your odds of winning the big jackpot may not be the greatest, but your odds of winning one of the smaller prizes get better with each draw.

So whether someone is lucky enough to win the lottery, score the ideal roll in Monopoly, or win a coin toss, remember that luck has nothing to do with it. So if you want a sure thing, be sure to check your odds and do the math. In the meantime, Keep Looking for the Science in Things.

Science In Pop Culture

Journey to Alpha Centauri: A Review of the Netflix Reboot – Lost in Space

(Originally Published April 20, 2018)

Interstellar space travel, once a dream at the far edges of our minds, is now slowly becoming a reality. When Lost in Space first aired on television in 1965, we knew very little about interstellar travel or what existed outside of our solar system, let alone on the outer fringes of our galaxy. As Netflix releases a new reboot of the television show, we have taken leaps and bounds closer to Lost in Space becoming a reality, in more ways than one.

The original premise of the Lost in Space series was a space-age Swiss family Robinson stranded on a strange planet due to a cataclysmic event while en route to Alpha Centauri, the closest star to our solar system. In the reboot, much stays the same in that they are still a space-age Swiss family Robinson that experiences a cataclysmic event that causes them to be stranded on an alien planet. However, the show has been updated with the recent science and knowledge that we have gained about the Alpha Centauri system.

Alpha Centauri is the closest star system to our solar system. In the past year or so, you may have heard a fair amount about Alpha Centauri, and in particular Proxima Centauri and Proxima Centauri b, in the news. Why is that? Well, Proxima Centauri b is a planet that orbits Proxima Centauri in the Alpha Centauri system. What makes Proxima Centauri b so noteworthy in recent news is that it was recently discovered to be orbiting within the habitable zone of Proxima Centauri (Emspak, 2016). This means that it is likely that water can exist in liquid form on Proxima Centauri b. And we all know that where there is water, there can be life.

So, with Earth in dire straits in the show, the Robinsons are part of many groups of colonists heading to another planet for humans to inhabit, in the Alpha Centauri system. But how far away is Alpha Centauri, and how long would it take or how fast would a spaceship have to travel?

Alpha Centauri is approximately 4.35 light years away, and Proxima Centauri is 4.25 light years away, from Earth (National Aeronautics and Space Administration: Goddard Space Flight Center, 2016). Contrary to popular belief, a light year is a unit of distance, rather than time. But how far is a light year?

$Light\ year = (speed\ of\ light)\ x\ (one\ year)$

$Speed\ of\ Light = c = 3\ x\ 10^8\ m/s$

$One\ year$
$= (365\ days\,/\,year)\ x\ (24\ hours\,/\,day)\ x\ (60\ minutes\,/\,hour)\ x\ (60\ seconds\,/\,minute)$
$= 3.15\ x\ 10^7\ s$

$1\ Light\ year = (3\ x\ 10^8\ m/s)\ x\ (3.15\ x\ 10^7\ s)$
$= 9.45\ x\ 10^{15}\ m$
$= 9.45\ x\ 10^{12}\ km$

So, knowing that we can figure out how far Alpha Centauri is from Earth in meters and kilometres:

$Alpha\ Centauri = 4.35\ light\ years = 4.11\ x\ 10^{16}\ m$
$= 4.11\ x\ 10^{13}\ km$

According to the latest version of Lost in Space, the colonists are said to have travelled about 26 trillion miles in order to reach Alpha Centauri, which is about 41,842,944,000,000 km, or $4.18\ x\ 10^{13}$ km, so that checks out. However, in episode 4, I have yet to find out how long they expected the journey to take. So if we assume the timeline of the original 1965 Lost in Space, the Robinsons expect it to take 5.5 years to get from Earth to Alpha Centauri.

$$5.5 \text{ years} = 2008 \text{ days} = 48,192 \text{ hours}$$
$$= 2.89 \times 10^6 \text{ minutes} = 1.73 \times 10^8 \text{ seconds}$$

So, how fast do the Robinsons need to travel to reach Alpha Centauri in 5.5 years?

$$\text{Speed} = (4.11 \times 10^{13} \text{ km} / 1.73 \times 10^8 \text{ seconds})$$
$$= 237,572 \text{ km/s} = 237,572,254 \text{ m/s}$$

Talk about breakneck speeds. It is less than the speed of light, so it's not an impossibility, according to special relativity. But how close are we to being able to travel to Proxima Centauri, like in Lost in Space?

The most recent mission to test the speed of an interstellar-capable vessel was the New Horizons mission, which launched in January of 2006. The top speed that the space probe reached was 45 km/s, which is the fastest achieved to date (Jackson, 2017). So how long would it take to get to Alpha Centauri at that speed?

$$\text{Trip duration} = (4.11 \times 10^{13} \text{ km}) / (45 \text{ km/s})$$
$$= 9.133 \times 10^{11} \text{ s} = 1.522 \times 10^{10} \text{ min} = 2.537 \times 10^8 \text{ hours}$$
$$= 1.057 \times 10^7 \text{ days} = \sim 28,962 \text{ years}$$

It doesn't look like that would work out too well for us. But! There is another craft with an even higher projected top speed that is scheduled to launch this July (2018), the Solar Probe Plus ("Parker Solar Probe," 2016). This solar probe is expected to reach a top speed of 200 km/s (Scharf, 2013). So how long would it take to reach Alpha Centauri?

$$\textit{Trip duration} = (4.11 \times 10^{13}\,km) / (200\,km/s)$$
$$= 2.055 \times 10^{11}\,s = 3.425 \times 10^{9}\,min = 5.708 \times 10^{7}\,hours$$
$$= 2.378 \times 10^{6}\,days = \sim 6{,}516\,years$$

Less than a quarter of the speed of the New Horizons', but still not a possible journey in a person's lifetime. However, if we can reduce the trip duration by that much in 12 years, what can we accomplish in the next 12 years?

All in all, the Lost in Space Reboot is a great family-friendly show (TV-PG rating), that does take recent developments in space discoveries into account, while staying true to some of the aspects that we were introduced to in the original, such as family relationships, teamwork, and valuing the strengths in others. It is currently on my Netflix list, and it should be on yours too. If you haven't checked out the Lost in Space reboot, be sure to do so. And don't forget! Keep Looking for the Science in Things.

George of the Urban Jungle: A Look At The Movie Rampage

(Originally Published April 27, 2018)

Giant albino gorillas, flying wolves, and monstrous alligators have recently been seen wreaking havoc on the silver screen in the new blockbuster, Rampage, starring Dwayne "the Rock" Johnson. But is it possible to manipulate an animal's DNA so that it can suddenly start growing, much like the gorilla George experienced? Today, we will be looking at how likely George the Gorilla is in the real world.

A male mountain gorilla averages about 6 feet tall standing fully erect ("GORILLA FACT SHEET," n.d.). Fun fact: Silverback gorillas are dominant male mountain gorillas that are 12 years old and older. In the movie Rampage, George comes into contact with a rock-like object from outer space, and once he touches it, he starts to grow to King Kong proportions. Now, the event of a magical or radioactive rock plummeting to Earth is quite unlikely. Don't get me wrong, space material, such as meteors and old satellites, fall towards Earth quite regularly; just none of it has had the power to genetically alter anything, and more often than not, it breaks up in our atmosphere and never reaches the ground. But what if we take the extraterrestrial rock out of the equation? Is it possible for someone in a lab to alter an animal, such as an adult gorilla, genetically?

All the research papers that I found on bioengineered and genetically modified animals mentioned that their test subjects were altered at the pre developmental or embryonic stages. What does this mean? Scientists introduce the snippet of DNA that they are interested in into the animals' DNA when they are only a collection of a few cells, or even before that when they are gametes (sperm and unfertilized egg cells). In the early stages of development, the single-cell fertilized egg continuously multiplies, copying its DNA repeatedly, until you have a multicellular embryo with each cell having the same DNA. For the altered organisms, each cell contains the snippet of introduced DNA as well. Which is much easier than having to try to override the trillions of cells, with the introduced DNA code, that the animal's body has at birth or as an adult.

But what could happen if we tried to introduce a new DNA strand into an adult animal? Well, in addition to having introduced DNA to many, many more cells, there are mechanisms in animals and our bodies that try to prevent changes to "the norm". First of all, it can be speculated that most individuals have very active immune systems that could view the introduction as a foreign body invasion and attack the DNA fragments as well as the altered cells ("Understanding the Immune System," 2003); thus, preventing any change from taking place and causing extensive damage. Also, if it manages to make it past the immune system, we have little "editors," known as RNA, that scan your DNA when a cell is going to replicate and try to repair any mistakes and changes to our DNA that could result in cell mutations, some of which could have negative consequences (e.g. becoming tumours) (Clancy, 2008 | "How cancer starts," 2017). As you can see, it is much simpler and more effective to introduce the alteration to the DNA at the beginning of the animal's cellular development, so the system is led to believe that it is always there.

That doesn't mean that there hasn't been work done on animals to try to create rapid growth. One recent example is AquAdvantage salmon, which has been shown in the lab to grow at twice the speed of their unaltered counterpart ("AquAdvantage Salmon Fact Sheet," n.d. | Moss, 2010). Although, in the end, they aren't any bigger than the salmon that takes the traditional amount of time to grow, scientists are hoping that by speeding up the growth rate, fish farms can produce the fish at a faster rate and get them to market sooner ("Technology," n.d.). How are they achieving this? Well, like I mentioned in a previous post, humans have two of every set of chromosomes; the salmon that are a part of this experiment have three of every set of their chromosomes. Scientists are replacing a snippet of DNA on one of the three chromosomes with a fragment from a fish that is known to grow very rapidly. This slight alteration prevents any major changes to the fish while just introducing characteristics of accelerated growth into its repertoire.

Now, I know you're probably thinking, but what about those people who are short one day and suddenly become giants? Well, there are three different instances that have been fairly well studied that can explain giants quite easily. The most common causes of extreme growth are gigantism, Marfan syndrome, and acromegaly. Much like humans, animals can also experience gigantism and other abnormalities due to disruption of the endocrine (hormone) system, just to a varying degree and, of course, not as extensively studied as in humans (Fero et al., 1996). Gigantism starts to occur during childhood when the individual gets a greater-than-normal dosage of growth hormones ("Gigantism," n.d.). This quite often occurs due to a tumour pressing on the pituitary gland. From this, we can rule that George did not experience gigantism as he is a mature adult gorilla. Marfan syndrome is a genetic disorder that is inherited from one or both of an individual's parents and affects the connective tissue, so it is not only characterised by abnormal growth but other irregularities, such as in the heart, lungs, eyes, etc. ("What Is Marfan Syndrome?" 2014). None of which are apparent in George, as well as none that were exhibited as a young gorilla. Now finally, acromegaly, which occurs after an individual reaches maturity, and is mostly due to a tumour pressing on the pituitary gland ("Acromegaly," 2012). Many speculate Adam Rainer had acromegaly ("Adam Rainer," 2017). However, it may also be a bit of a stretch to say that George had acromegaly, as none of his other features changed like they do for individuals that have it (changes in facial structure).

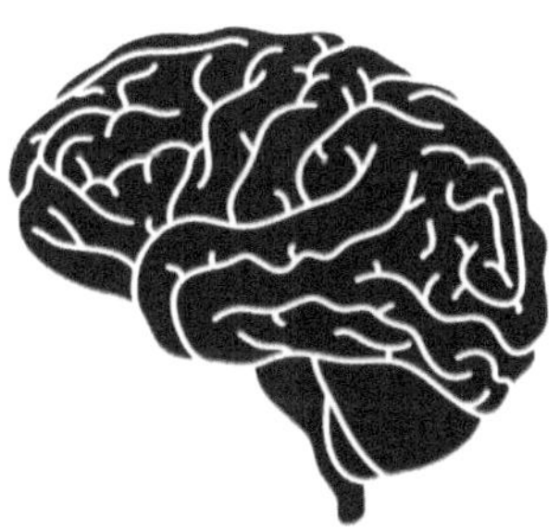

Ok, so what you are trying to say is that George is 100% impossible, aren't you? Well, the accelerated growth is unlikely without other symptoms and changes occurring either. However, prior to his contact with the "magic" rock from space, George as an albino male gorilla was possible, although rare. Albinism occurs through the mutation of the gene that codes for pigment (e.g. skin pigment, iris pigment, etc.) (Ballantyne, 2009). In fact, there is only one known documented case of albinism in gorillas, and it was a gorilla by the name of Snowflake, who was housed at the Barcelona Zoo until his death in 2003 from skin cancer, which albinos are at a greater risk of contracting (Didymus,

2013). In addition to being more prone to skin cancer, albinos are also more sensitive to light, which would have had George sticking more to the shade rather than charging out in broad daylight (Miller, 2010).

Despite being based on a video game, Rampage brings forth questions about the capability of bioengineering as well as natural occurrences. Although it is not possible for an adult gorilla to suddenly experience an extreme growth spurt without some underlying medical condition, it is not unlikely that we may see supersized animals in the future with the way that technology and innovation are headed. Just not how it plays out for George in the movie. It's also interesting to see that rather than just another King Kong lumbering across the screen, we have an opportunity to see and discuss the rare natural occurrence of albinism.

Thanks for tuning in to today's post. It was definitely interesting to try to find some of the science behind Rampage. Until Monday, Keep Looking for the Science in Things.

Technology Behind Avengers: Infinity War

(Originally Published May 4, 2018)

As many may know and many may have already seen, last Friday we were given the next, long-awaited installment of the Marvel Universe. Avengers: Infinity War. Now, I know no one likes spoilers, so I won't ruin the movie for those who haven't had the opportunity to go see it. However, whether you have seen the movie or not, I am going to discuss some of the technology of one of our favourite Marvel heroes that has been seen both in this new movie and in their own self-titled movies. This individual is known for their innovation, technology, and gadgets that they have created. They have contributed so much to the Marvel world of technology that we are only going to have a chance to highlight some key elements in today's post (stay tuned for further discussion in future posts). If you haven't already figured out who I am talking about, well, today we are going to discuss some of Tony Stark's technology, otherwise known as Iron Man.

Arc Reactor

When we think of Iron Man, we know that he wouldn't be Iron Man without his arc reactor. Even his suit is secondary to the arc reactor. For those that aren't familiar with Tony's arc reactor, it is a device that is implanted into Tony's chest to prevent metal shrapnel that is embedded in his chest from hitting his heart. Now there are certain aspects of the science behind the arc reactor that are still in development (on a much larger scale), and there have only been speculations on how Tony's arc reactor would work. One of the most in-depth looks at Tony Stark's arc reactor comes from the blogger Gizmodo (Carlyle, 2014). From what Gizmodo has gathered, the fuel source of the arc reactor is a palladium alloy made up of palladium-103 and palladium-107, which has some interesting decay properties that may have been harnessed in the Marvel Universe but are not a possibility in the real world.

Palladium is a transitional metal that has an atomic number of 46. This means that it has 46 protons (with a positive charge), 46 neutrons (with a neutral charge), and 46 electrons (with a negative charge), with no net charge. Now, palladium-103 still has 46 protons and electrons, but it has 57 neutrons, and palladium-107 has 46 protons, 46 electrons, and 61 neutrons. Although they are more stable than some of palladium's other isotopes, they want to reach a more stable state, and in order to do so, they must undergo radioactive decay. It's because of this desire to be more stable that palladium has been used in cold fusion experiments, which are similar to regular fusion but do not require hot plasma or containment (e.g. cooling ponds). However, there's an issue with real-world science. When palladium-103 (Pd-103) undergoes radioactive decay, it performs an electron capture, which is where the inner electron is absorbed by the nucleus of the atom, merging with a proton to become a neutron, which produces a burst of energy (a gamma ray); thus transforming into a new element Rhodium-103, which is much more stable. Now when palladium-107 (Pd-107) undergoes radioactive decay, it undergoes what is called beta decay, which is when it ejects an electron when a neutron changes into a proton, and becomes more stable as Silver-107.

Now in the real world, as these elements decay, there is no net electric charge to power the electromagnet, stopping the shrapnel. So, Gizmodo proposes that Howard Stark (Tony's dad) discovered a way "using comic-book physics" to use Palladium-107 as the electron source for Palladium-103, thus establishing a circuit and the basic gist of how it works. I definitely recommend reading Gizmodo's article, as he really does go into depth and shows proof of the arc reactor being a palladium alloy (*warning* article is rich in science jargon).

J.A.R.V.I.S.

If you have seen any of the Iron Man movies or any of the Avengers, you have met Tony Stark's right-hand man, J.A.R.V.I.S. Just A Rather Very Intelligent System (J.A.R.V.I.S.) is an artificial intelligence that essentially manages Tony Stark's life for him. From controlling Tony's home settings and security to medically diagnosing Tony, monitoring the status of the Iron Man suit and conducting updates to the software, managing and conducting all of Tony's computer tasks (saving files, transferring files, etc.), relaying phone calls, pinpointing targets, and advising Tony Stark on everything from mission objectives to everyday life. It is safe to say that without J.A.R.V.I.S., Iron Man wouldn't be who we know him to be. And I know that I am one of many who would love to get their hands on their own personal J.A.R.V.I.S. So when can we expect one in our own home? Well, there are already basic prototypes available now.

One of the more expensive and extensive models is owned by none other than Mr. Facebook himself, Mark Zuckerberg (Woollaston, 2016). Unveiled at the end of 2016, Zuckerberg's model showed that it was capable of controlling his home (which includes his lighting, temperature, appliances, music, and security), as well as having a learning capability as well. However, it is not quite as humanoid as J.A.R.V.I.S., and we may be a ways off until that happens.

Now, I don't know about you, but Mark Zuckerberg and I are very far from being in the same tax bracket, let alone having the same level of funding for my own J.A.R.V.I.S. Well, we may not be out of luck. Two young engineers, Shubhro Saha and Charlie Marsh, have developed an open-source, fully customizable platform that allows you to make your own J.A.R.V.I.S. out of cheap, off-the-shelf computer parts and the free software available on their site. It goes by the name of Jasper (link to the Jasper Project in the Works Cited).

Not a computer tech? Well, rather than calling your system Jarvis or Jasper, you may not need to look any further than Alexa and Google. Recently, companies such as Amazon and Google have come out with their own consumer-friendly equivalent of J.A.R.V.I.S., which can be asked the weather, the time, control already installed "smart" home technology (lighting, temperature), play music, add things to your grocery list, make cat noises, and tell jokes. However limited they currently may be (they can only perform tasks that have been programmed into their software and aren't learning), there may be models and updates in the future that expand their capability to that of J.A.R.V.I.S.

Iron Man Suit

I'm sure everyone can agree that one of the coolest pieces of tech that Tony Stark has created is the Iron Man suit. Tony's exoskeleton provides him with superhuman strength, flight, a multitude of weapons, and has various sensors and means of communication. Not only do many people in the Marvel Universe want to get their hands on their own Iron Man suit, but many governments, corporations, and even private citizens have set about making their own equivalents.

SOCOM is a special branch of the United States military that is currently developing Iron Man-inspired suits, which they are calling Tactical Assault Light Operator Suits (TALOS), which they are expecting to release in August 2018 (last reported date) (Medina, 2015). Through these suits, they hope to increase a soldier's strength, mobility, and safety while in combat. One special feature of their suit (no, it's not flight) is that it has a liquid body armour that solidifies when a magnetic or electric current is applied by the wearer through a switch, making the armour highly resistant to fired ammunition. They hope to equip both Marines and Navy Seals with this style of armour upon release.

In the future, you may just see Iron Man assembling your next family automobile. Well, not really, but even the car manufacturing companies are looking into aspects of Iron Man's suit to help improve job site safety and the health of their employees (Tegler, 2017). Many employees that work in assembly plants have to deal with highly repetitive tasks and motions, which can take a toll on their bodies. So many manufacturing companies are looking at using exoskeleton assists for their employees. These suits are unpowered, so they do not amplify their strength but rather assist in tasks that they are already able to conduct. Such as supporting a worker's arms when they do repetitive tasks about their head, so the exoskeleton takes the strain rather than the worker's shoulders and upper body. Another example is that rather than a worker crouching, the exoskeleton supports the worker's weight and allows them to be in a seated position, with the exoskeleton taking up the strain that would otherwise be on the worker's joints.

Most examples of the Iron Man suits that have been created by others have failed to replicate one key characteristic of Tony Stark's Iron Man suit, and that is the ability to fly. However, one man has harnessed the ability to fly, and that man is Richard Browning (Dunning, 2017). Still, in the prototype phase, Browning can lift himself off the ground by using six jets that are attached to his forearms and his back. If that sounds like a lot of physical strength in order to control them, you'd be right. Browning says that his prototype is capable of "pushing 130 kilos of thrust across [his] entire body". And his workout routine reflects that. So it appears that the average human will have to wait until flight stabilizers are developed before we will be able to fly like Iron Man.

Well, I hope you enjoyed this look at the science and technology behind Iron Man. There is definitely much more to dive into when it comes to the innovations of Tony Stark, but that will have to be something saved for a future post. And remember, keep Looking for the Science in Things.

Finding Mermaids: Analyzing Mermaids: A Body Found

(Originally Published May 7, 2018)

I still remember the day when I was misled by a source that I thought I could trust. I was around 11 or 12, and I was watching what was marketed by Animal Planet as a documentary about dragons. The Last Dragon. For the full hour and a half, I believed that they had found the remains of a dragon frozen in ice in a cave in Romania. I excitedly told my mom about it - about the creature that had been a part of my imagination, along with the unicorns and fairies - and how it suddenly was a reality. It had been real. That was until the program ended, and it was followed up with how they made it. And just like that, the dream shattered like a crystal chandelier crashing onto a marble floor. I felt embarrassed that I had fallen for the sham and that I had told my mom with such excitement. That's when I began to feel that belief was not enough anymore. I needed proof. I couldn't just blindly trust. Well, ladies and gents, it seems that Animal Planet has done it again. They've released a production about mermaids. Mermaids: A Body Found. And not only released but reaired it. So, I think it's only fitting that we see if this is a documentary or a mockumentary.

Brought To You By Animal Planet

The first thing that leads us to believe that we are about to watch a documentary is its classification on many media sources. I found the documentary listed on iTunes as non-fiction. That would imply that it is factual and not a work of fiction. In addition, it was broadcast on Animal Planet, which is known for its factual animal documentaries.

Largest Mass Whale Beaching in US History

Quite quickly, the scene is set for us. April 4, 2004. In Moclips Beach, Washington State. We are told that there has been a whale beaching, but not just any beaching, but the largest mass whale beaching in US history. So surely, I must be able to find something about this whale beaching event. An extensive search through Google for "April 4, 2004 whale beaching," "2004 Washington state whale beaching," and "largest mass whale beaching in US history." Surprisingly, nothing turned up in the state of Washington. So I turned to the archive at the Washington State Library, and again, nothing turned up on April 4, 2004, or Moclips Beach. In fact, the largest mass whale beaching in US history happened in Hanalei Bay, Kaua'i, on July 3-4, 2004, when between 150 and 200 melon-headed whales were beached, with sonar activities being questioned as the cause (no biological, oceanographic, or meteorological reasons were found, and there were sonar activities being conducted on July 3, 2004) (*NOAA Magazine,* 2006).

Checking The Credentials

Of all the things about this film, this next one made the hairs on the back of my neck stand up on end. For the duration, we are introduced to different individuals claiming to be scientists. A quick Google search and comparison of photos reveals that these individuals are just actors. Even in the end credits, and disclaimer after the end credits they do not mention these false titles, or they are merely actors. Now, why does this bug me? Well, I may only have my Bachelor of Science, but I worked for five long years to get those credentials. To have the right to have BSc after my name. Now the actors are claiming that they are doctors and are leading people to believe that they are such and that their expertise can be trusted. This lie hurts actual scientists who have spent many years of their lives becoming experts in their field and trusted sources (the minimum schooling to have your doctorate in science is seven years, depending on the program).

Trusting An Insider

In addition to the scientists, we are introduced to an "Insider," an individual that had been with the US Navy and was on the beach on April 4, 2004, who saw the body. Of course, when divulging secure information, it is expected that they are blacked out and that their voice is altered. However, this insider is not given a pseudonym or fake name; his rank is not mentioned, nor are his years of service given. In other documentaries, such as the ones that feature undercover cops, they are given a rank and years of service - some piece of non-specific identity that gives them some credibility. This is not the case in Mermaids: A Body Found. We must trust everything he says.

Proof In The "Actual Recording"

Throughout the film, we are flashed between re-enactments, interviews, a simulation of the evolution of mermaids, back to a re-enactment, presenting a piece of evidence, and then into an interview. Sometimes, when we flash to a re-enactment, we see the words flash across the screen, "dramatic re-enactment." Other times, it does not. After a while, it gets quite confusing as it blurs the lines between what may be fact and what may be fiction. But if you can keep it straight, you will notice that most, if not all, of the evidence for mermaids is presented during a re-enactment sequence. In addition, most of the tangible evidence that they had was either destroyed or confiscated. So it cannot be proven whether it is real or not; we can't even test the tissue sample because it was destroyed. The only true evidence that we are left with is the "actual recordings," one from Washington and the other from South Africa. It is claimed that the recordings are public records, so that means that the public would have access to them. Another search of Google did provide the recordings, but nothing prior to the film's release (which would be available if they were, in fact, real), and there was only one recording, not two, as there should be. So I went back to the film to take a closer look at both recordings. Interestingly enough, there were never two recordings. The audio and imagery were the same for both. The communication between mermaids and dolphins that was pointed out in the second recording was present in the first recording; it was just never pointed out.

What Is The Bloop?

There is mention of "the Bloop" or an audio recording having a "bloop signature". But what exactly is the Bloop? We are told in Mermaids that it was recorded by NOAA in 1997, and we don't know what caused it or made that sound. However, one thing that I noticed was that the Bloop was not actually played. So when it is mentioned that something is like the Bloop or has a Bloop signature, we actually have nothing to compare it to. Mhmmm... That's interesting. Why would they omit that crucial piece of evidence from reality? Well, it's because the recordings that have been played for us sound nothing like the Bloop or have a "Bloop signature" (I've included a link to the Bloop in the Works Cited for you to listen for yourself).

Evidence In Writing... Or Painting

In the film, they highlight a cave painting from a cave in Egypt that depicts mermaids. No location is given, but again, a quick Google search and I found the Cave of Swimmers ("Cave of Swimmers, Egypt," n.d.). I was a little taken aback as I saw the picture and thought I must be mistaken. There must be another cave, but alas, there is none. The film added to the mural and definitely embellished it a fair bit.

Disclaimers Ahead

And finally, at the end of the film, there are several disclaimers. Where? After the end credits, there are a few disclaimers that flash on the screen for a matter of seconds, and you have to pause it in order to get enough time to read them. "None of the institutions or agencies that appear in the film are affiliated or associated with it in any way, nor have approved its contents. Any similarities in the film to actual persons, living or dead, is entirely coincidental." "Though certain events in this film are fictional, Navy sonar tests have been directly implicated in whale beachings." There you have it, ladies and gentlemen, fiction in its deceptive form. They still don't clarify what was fact or fiction, but I'm sure you have been able to figure it out. But still, there should be more than just a single disclaimer at the end of the credits (often cut out by TV broadcasters).

So Is Any Of It True?

There's a statement early on that in the early 2000s, it was proven that whales beached themselves during Navy sonar tests (Nieukirk et al., 2004). And it's true. Over the past decade, more and more studies have come out stating the effects of sonar on the echolocation abilities of whales (Fernández et al., 2016). In many cases, sonar results in confusion for the whales, resulting in them beaching themselves. And in many cases, the whales pass away because they can't get back out to deep water in time. They have even extended it to noise pollution in general, including offshore drilling rigs (Malakoff, 2001).

So Why Should We Watch This Film?

Because of what it represents. Mermaids: A Body Found is more than just a mockumentary about mermaids. It's a protest that uses a shock strategy to bring it back to the forefront of our attention. It is true that whales are being beached due to sonar testing and various noise pollution sources, including underwater drilling. Would the film have grabbed your attention if it was about whales instead of mermaids? If we look at what the actor Dave Evans says at the end of the film, "I started thinking of them as fellow beings. And the last time we lived alongside a fellow human species was with Neanderthals. We displaced them. Some scientists think we actively hunted them, that we devoured their existence. It's the same thing. What we did on land to one of our own, we'll do in the sea. We're not good at coexistence. Sometimes other responsibilities supersede science. My only goal is that the Navy stops their sonar testing, that court inquiries open into their activities that then lead to court orders to stop it. Mermaids have persisted only because they can hide. And I hope they stay that way. I hope they stay hidden. I don't want to hunt them anymore because... they don't want to be found." Now replace mermaids with whales. Do you still care?

Genetic Memory: Looking at Assassin's Creed

(Originally Published May 18, 2018)

With 20 games spanning 16 different platforms, bringing to
life Ptolemaic Egypt, the Third Crusade, the Italian Renaissance, the
Colonial Era, the French Revolution, Imperial China, the Victorian
Era, the Sikh Empire, and the October Revolution, Assassin's Creed is
one of the bestselling and most highly played game franchises currently
on the market. Started over a decade ago, Assassin's Creed has since
broken out of the video game world and ventured into comic books,
books, a major motion picture in 2016, and current talks of a TV series.
For those who are not familiar with the franchise, here's a quick
summary: It is based on a long-standing feud between the Assassins and
the Knights Templar. As a player, you control a descendant of one of the
Assassins, and at times the assassin themselves, who is currently being
used by a corporation owned by the Knights Templar to find
information about the location of historical artifacts that were hidden
by the Assassins from the Knights Templar. How do they extract this
information? Through a piece of machinery called the Animus, which
pulls the character's ancestor's memories out of the character's DNA
and makes the character relive these memories. This may seem like a
far-fetched notion; however, we may be closer to that reality than we
know.

In biology, there is a field of study called cellular genetics, which involves
the study of memories being recorded within genetic material and then
being able to be inherited by the individual's descendants. Scientists
have discovered a few instances where this may actually occur.

One example of this is viral DNA. In order for viruses to reproduce,
they utilize the host's DNA to replicate themselves. In many cases,
the virus leaves the host cell intact and, with it, a segment of viral DNA
code in the host's DNA, even after the host has fought off the virus
("Virus Infections and Hosts," n.d.). This segment is often passed
down to the host's descendants, and geneticists can pick out these viral

markers many generations later (Zimmer, 2015). In addition, although the descendant may not have experienced the virus, it provides the descendant's immune system with a "guide" of sorts on how to defend the body from that virus, making the descendant immune or not affected as greatly as their ancestor was.

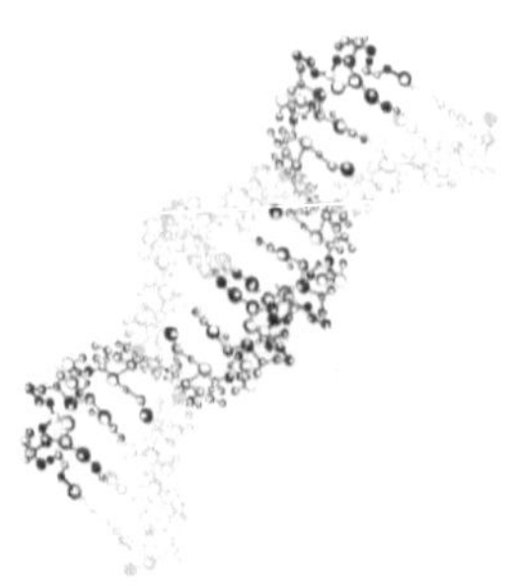

Another example of inherited memories actually falls in the field of animal behaviour and epigenetics. Epigenetics is the study of how a living being develops and functions due to forces and influences other than its DNA (Walia, 2015). Recent studies have concluded that it is possible for a subject to pass an avoidance behaviour on to their descendants. In one study, mice were taught to fear the smell of cherry blossoms (Dias & Ressler, 2014). The following two generations were shown to have a higher sensitivity to the smell of cherry blossom, and they continued to avoid the smell despite never having encountered it in their lives (Gallagher, 2013). It was discovered that the fear of the cherry blossom smell was in fact transferred genetically. And it could go on to provide explanations for people's phobias or abnormal behaviours in specific situations (Walia, 2015). This being a fairly young field of study, it will prove interesting in the years to come to see what other information we inherit from our ancestors. And to see if complete details of memories can be extracted from the DNA of someone who has not directly experienced them. Now what about the Animus? Could we become someone's lab rat for the purpose of extracting information that they want?

In the game, when an individual is hooked up to Animus, they enter the memory like a dream, which they can interact with. This view is then

projected onto a screen for the Animus technician to record and view.
As space-age as this idea appears to be, scientists today are able to project
images from a person's mind. Researchers at UC Berkeley were able to
decode visual information in the brain and apply it to their computer
algorithm, which then recreated the image based on what the brain said
it had seen (Alvarado, 2012). In order to do so, they had their test
subjects watch movies while they attempted to translate what the brain
was "saying" it was viewing (Nishimoto et al., 2011). Although the
images are not very clear and may vary in some aspects, the similarities
are remarkable.

So it appears that we may be quite close to being able to replicate
Assassin's Creed, except for a few issues. We have yet to find a way to
isolate exact events and details from an inherited memory. Any image
projected from the mind, although possible, appears quite blurry and
may vary from the true image. And finally, the images we project from a
person's mind are their experience, generally in real time, rarely a
memory, and always of that person. So don't worry, you won't be
getting called in as a key witness to a crime that your great-great-great
grandfather witnessed any time soon. Nor will you be forced to give out
secret information that you didn't even know you had. Thanks for
reading, and remember: Keep Looking For the Science in Things.

Regrowing Deadpool

(Originally Published May 25, 2018)

Last Friday, thousands of people packed into theatres to see the much-anticipated sequel of the anti-hero Deadpool . And many have yet to see it, so today's post on Deadpool 2 will contain no spoilers. I promise. Instead, we are going to talk about an interesting phenomenon that is characteristic of Deadpool, and it is not his dry wit. But rather limb regeneration. In the first film, we see Deadpool regrow his hand, with many jokes at the expense of Al. And in Deadpool 2 , we see Wade Wilson up to his usual antics again. But is limb/body part regeneration even possible? Could people be capable of regrowing a lost arm or leg in the future?

Regeneration in Nature

Limb regeneration is not a stretch of the imagination. There are plenty of animals that are capable of regrowing body parts. Geckos drop their tails in order to escape predation and then regrow a new tail. Sea cucumbers regurgitate their internal organs as a way to get away from predators and then regrow the lost internal organs. And salamanders are capable of regrowing lost limbs and are the main subject of countless studies exploring limb regeneration. What makes these animals so different from humans? When a salamander loses a limb, a few key factors come together in order to regenerate the missing limb: the protein nAG, macrophages, and stem cells. nAG is a key protein that scientists have discovered stimulates the stem cells at the site of the amputation to regrow the lost limb (Bryner, 2007). Macrophages are a part of the immune system and are even found in our bodies as well as in other mammals. They produce molecules that signal other types of cells to grow the new limb (in salamanders) or heal a new wound (in mammals) (Lewis, 2013). One of these types of cells includes stem cells. However, in humans, many of these abilities, although genetically speculated to be there, are turned off (Emspak, 2017). Researchers are currently trying to find ways to turn this ability back on.

Stem Cells

One main area of focus in regenerative studies is stem cells. Embryonic, unspecified, stem cells are special because they are able to become any specialized tissues and structures in the body ("Stem Cells Basics I." n.d.). They are the first type of cells that we possess during development. In many animals' development, embryonic stem cells go on to become skin cells, liver cells, nerve cells, eye cells, etc. Once development has been achieved, stem cells are still present in many body systems and tissues in order to aid in repair and regeneration. Some animals, such as salamanders, can regenerate full limbs from a set of stem cells. But when it comes to adult humans, there is a catch. Stem cells are still found in many of our systems and tissues (but not all), and they still regenerate some of our tissues. However, they can only regenerate the cells of the system or tissue they are a part of ("Adult Stem Cells," n.d.). For instance, bone marrow stem cells can only regenerate and replenish new bone marrow; they cannot be reallocated by the body to repair the spinal cord, which does not have any stem cells. This is why finding ways to find and utilize unspecified stem cells (like those found in embryos, not adults) is a major component of spinal cord injury research, as we are unable to regrow existing spinal cord cells. Yet, if we had access to unspecified stem cells, we could possibly regrow new spinal cord cells to replace the damaged ones. Despite this hurdle, in recent studies, scientists have been capable of growing many different body parts thanks to these special types of cells.

Petri Dish Body Parts?

Although there have been no cases of regrowing an arm or leg in a lab, a very current field of research in the medical sciences is the ability to regrow body parts and organs. Often called 'petri dish organoids,' these grown organs strive to one day shorten transplant lists and make black market organ harvesting obsolete, among other things. Although we are a ways away from being able to grow a full adult kidney in the lab, scientists have managed to grow fallopian tubes, a 5-week-old fetus brain structure, a mini heart-like structure, a mini kidney-like structure, a mini lung-like structure, and a mini stomach-like structure, among other structures that were replicated and transplanted into animal test subjects (Kessler et al., 2015 | "Scientist," 2015 | Ma et al., 2015 | Minoru et al., 2015 | Dye et al., 2015 | Geggel, 2014). As for human success in 2014, there was a study that was deemed successful. The scientists managed to grow and successfully transplant vaginas into four subjects due to birth defects: either the patients were born without one or they were underdeveloped (Gholipour, 2014). So it is quite possible that in the near future, there will be life-saving organ transplants with organs that were grown in a lab.

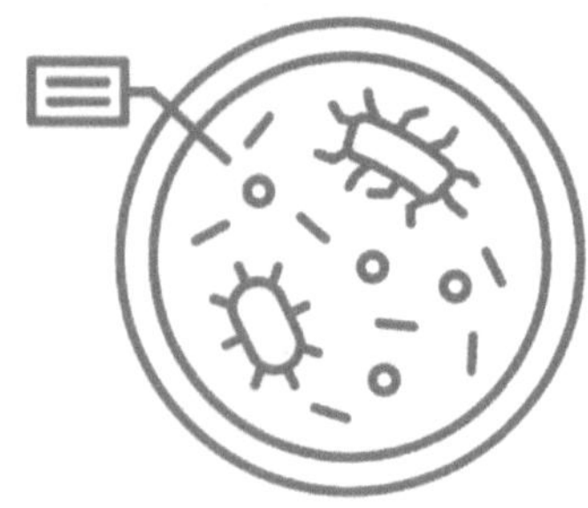

One of the Human Body's Superpower

Despite being unable to regrow a full leg, the human body is currently capable of some regeneration abilities. Skin is capable of regenerating several times over, as we have all experienced cuts and scrapes only to see them disappear without a trace (Ericson, 2013). Not quite as fast as Wolverine or Deadpool, though. Thus resulting in procedures such as skin grafting. Other tissues and structures that we regenerate but do not see include blood vessels, bones, and gut lining ("Valvular and Vascular Repair and Regeneration," n.d. | Dimitriou et al., 2011 | Okamoto, 2011). Possibly one of the most amazing is our liver. Although we cannot donate our entire liver and have it grow back, we can donate a lobe, have it grow back, and have the donated portion grow into a fully functioning liver within the recipient ("Liver Regeneration," n.d.). There are even reported cases of individuals who have cut the tips of their fingers off and are capable of regrowing (so long as the nail bed is intact) (Illingworth, 1974).

So, regeneration in the human body is possible in some aspects. Not quite to the extent that Deadpool/Wade Wilson is capable of, but scientists are optimistic that they may be able to unlock that ability in the near future. Until then, don't stick your arms or legs anywhere you shouldn't be, because you only get one set. For now. In the meantime, Keep Looking for the Science in Things.

Racing the Millennium Falcon

(Originally Published June 1, 2018)

Last Friday, we were given the next installment in the Star Wars franchise and the backstory to Han Solo and Chewbacca's friendship and lives. Not only that, but we were given further details on the iconic Millennium Falcon. Debated since the beginning of the franchise as one of the fastest starships in the galaxy, the Millennium Falcon has never been given a definitive top speed. But we may be one step closer.

According to Wookieepedia, the Millennium Falcon has a top speed of 1,050 km/h in an atmosphere and an acceleration factor of 3,000 G in space. So, unless it is traveling in an atmosphere or wants to know how hard it can accelerate, those won't help us that much in determining top speed. Our next tidbit of information comes from Han Solo himself, as he stated that the Millennium Falcon had completed the Kessel Run in less than 12 parsecs. It sounds great until we find out that a parsec is a unit of distance (31 trillion kilometers or 3.26 lightyears), not a unit of speed or time. A final tidbit that Solo gives us is, "makes .5 past light speed." Without knowing what .5 represents, all we can gather from this statement is that it will go faster than light speed. There are a few websites that have tried to tackle this problem before, all with varying results, but the best model to follow is that of Chris Lough of TOR.com from 2014.

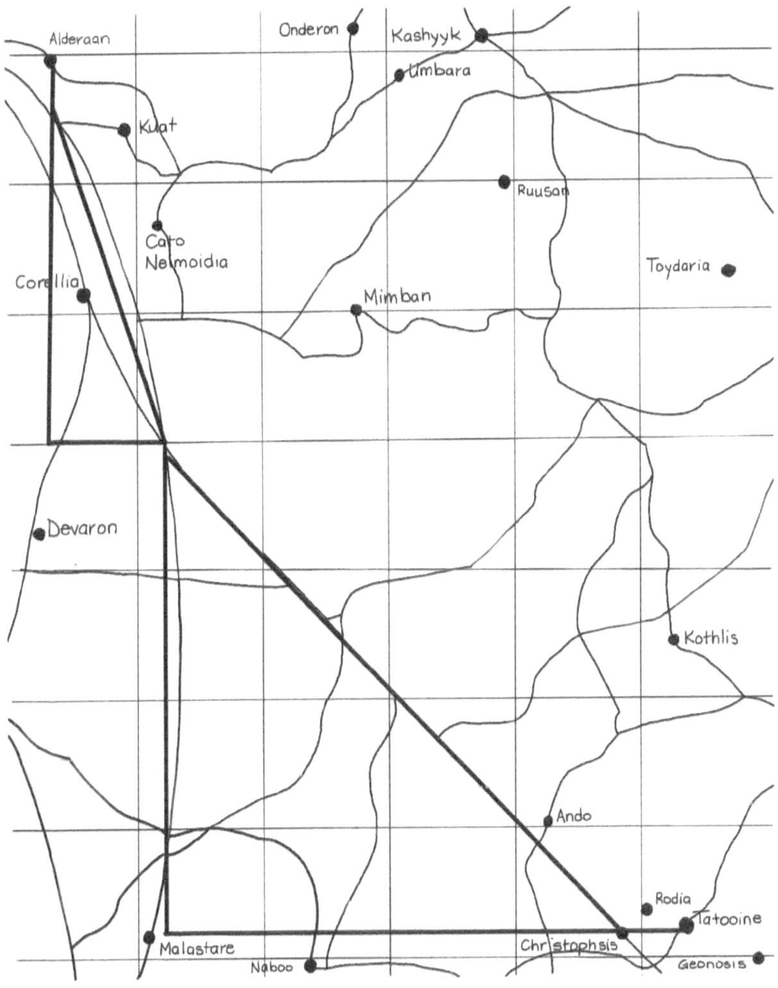

The best maps of the Star Wars galaxy come from those that are based on The Essential Atlas: Star Wars, which has been proven up until today to be the most accurate representation of the Star Wars galaxy. So I have drawn my own representation of that map, and that is what we will be using to make our calculations. There is no scale on the map, so it appears that we need to find the size of the galaxy. Using Wookieepedia, as our credible Star Wars source, they have the galaxy listed as being over 100,000 lightyears (as of May 30, 2018). The map is a 21 x 23 grid, to determine the distance of each square:

As we discussed in our Lost in Space post, a lightyear is a unit of distance, not time, as many are led to believe. One lightyear is equal to 9,450,000,000,000,000 meters (or 9,450,000,000,000 km). So the distance of one square on the map is equal to 4.11×10^{18} meters (4.11×10^{15} kilometers). Now, to figure out how fast the Millennium Falcon can go.

One scene from the movies that we can use to determine speed is from Star Wars: A New Hope, of the trip from Tatooine to Alderaan. This scene is useful as it does show some passage of time, both on and off the Millennium Falcon. So let's take a look at the distance they travelled.

Without a map scale, we cannot directly determine the distance that the Millennium Falcon travelled. However, we do know that each square length equals 4,347.82 lightyears. Now, thanks to the power of my computer's Paint program, we can see that the route is roughly two right-angle triangles.

Now how many people were wondering in high school "When will I ever use Pythagoras theorem again?" Well, we are right now. In case you don't remember, Pythagoras' Theorem is that equation that goes $a^2 + b^2 = c^2$, with c^2 being the hypotenuse or the length opposite the right angle (the longest side). So starting with the biggest triangle first,

$$a^2 + b^2 = c^2$$

$$4 \text{ squares}^2 + 4 \text{ squares}^2 = c^2$$
$$16 \text{ squares} + 16 \text{ squares} = c^2$$
$$32 \text{ squares} = c^2$$
$$5.66 \text{ squares} = c$$

In other words, c = 5.66 x 4,347.82 lightyears = 24,608.66 lightyears

Now for the smaller triangle,

$$a^2 + b^2 = c^2$$

$$3 \text{ squares}^2 + 1 \text{ square}^2 = c^2$$
$$9 \text{ squares} + 1 \text{ square} = c^2$$
$$10 \text{ squares} = c^2$$
$$3.16 = c$$

So, c = 3.16 x 4,347.82 lightyears = 13,739.11 lightyears

For a grand total of 38,347.77 lightyears traveled in total. So now that we know how far they travelled, we need to figure out how fast they were going. If you haven't seen Star Wars: A New Hope, I'm sorry there are spoilers ahead. Also, if you haven't seen the movie yet, what have you been doing? The movie has been out for 41 years! You get no sympathy from me for wrecking the movie for you. Anyways… From watching Star Wars: A New Hope, we see that Leia was on her way back to Alderaan (appearing to be within sight of the planet) when she was captured by Darth Vader. For over half the movie, Leia appears to be in Darth Vader's custody, so by using a continuous storyline (no flashbacks or fast forwards), it would appear that she is in Darth Vader's custody for a minimum of several days, if not a matter of a week or two, fairly close to Alderaan's location. Also, in the film, we see that while Han Solo, Luke, Obi-Wan, and the others are on their way to Alderaan, they have time to play a game that resembles chess and Luke is able to fit in some lightsaber training. This indicates to the viewer that we are looking at a time frame of hours and days for the journey. With that in mind, we can create a speed bracket at which the Millennium Falcon is capable of travelling.

Hours Spent Travelling	Lightyears/hour	Kilometers/hour	Meters/second
2	19173.885	1.81E+17	5.03E+16
4	9586.9425	9.06E+16	2.52E+16
6	6391.295	6.04E+16	1.68E+16
8	4793.47125	4.53E+16	1.26E+16
10	3834.777	3.62E+16	1.01E+16
12	3195.6475	3.02E+16	8.39E+15
14	2739.12642857143	2.59E+16	7.19E+15
16	2396.735625	2.26E+16	6.29E+15
18	2130.43166666667	2.01E+16	5.59E+15
20	1917.3885	1.81E+16	5.03E+15
22	1743.08045454545	1.65E+16	4.58E+15
24	1597.82375	1.51E+16	4.19E+15
48	798.911875	7.55E+15	2.10E+15
72	532.607916666667	5.03E+15	1.40E+15
96	399.4559375	3.77E+15	1.05E+15

3.77E+15 is the same as saying 3.77×10^{15}.

Assuming the journey didn't take less than two hours (because who would pull out a game of chess for anything less than four hours) and no more than four days, it can be assumed that the Millennium Falcon's top speed is somewhere between 399 and 19174 lightyears/hour or 3.77×10^{15} to 1.81×10^{17} km/h (1.05×10^{15} to 5.03×10^{16} m/s). Although these numbers may seem fairly vague, we can compare them to some figures that we do know. The New Horizons Probe, launched in January 2006 by NASA, has been measured to reach a top speed of 45,000 m/s. And the Parker Solar Probe Plus, which is scheduled to launch later this year, is expected to reach a top speed of 200,000 m/s. So if these vessels were to take the same journey, this is how long it would take:

38,347.77 lightyears = 3.62 x 10^{20} km = 3.62 x 10^{23} m

New Horizons Probe

Time = 3.62 x 10^{23} m / 45,000 m/s
= 8,044,444,444,444,444,444.4 seconds
= 134074074074074074.1 minutes
= 2234567901234567.9 hours
= 93106995884773.7 days
= 255,087,659,958.3 years

Parker Solar Probe Plus

Time = 3.62 x 10^{23} m / 200,000 m/s
= 1810000000000000000 seconds
= 30166666666666666.7 minutes
= 502777777777777.8 hours
= 20949074074074.1 days
= 57,394,723,490.6 years

Needless to say, as much as everyone thinks the Millennium Falcon is a hunk of junk, it is still light years ahead of our latest technology. And that's all for today's post. In the meantime, Keep Looking For the Science in Things.

The Jurassic Park Question: 5 Extinct Animals Being Brought Back

(Originally Published July 10, 2018)

In one of our previous posts, we discussed animals that were on the brink of extinction that we have saved through human intervention. But what about those that have already become extinct? Can we bring them back to life and have these animals return to the wild, or will we live out the equivalent of Jurassic Park? Whether it's human guilt or genuine scientific curiosity fuelling these endeavours, we have to ask ourselves, should we? And to help us decide this, let's look at five de-extinction projects that are getting very close to accomplishing their goals.

Mammoth

This one is so widely discussed that I would be surprised if you hadn't heard about scientists looking into bringing back the woolly mammoth. As many of us know, mammoths went extinct around 10,000 years ago - well mostly. The reason I say mostly is that scientists recently discovered the remains of a population on Wrangel Island, a giant nature reserve off the coast of Russia in the Arctic Ocean (Lewis, 2015). Now, what makes these remains notable? Rather than dying out 10,000 years ago, they continued to persist for another 6,400 years. Also, many of the individuals exist in near-pristine condition, frozen in the ice. And with that, DNA has been extracted from these individuals. Interestingly enough, scientists have been able to determine from this extracted DNA that this final population was quite inbred due to their thousands of years in isolation (Lewis, 2015). However, with other specimens, such as Lyuba and Dima, superbly preserved infant mammoths, scientists hope to splice the recovered mammoth genes into elephant embryos to create mammoth-elephant hybrids that may once again roam Earth (Mueller, 2009).

Dodo

Another well-known extinct animal, the dodo bird, went extinct in 1662 due to hunting and the introduction of foreign species, e.g. pigs, to the island of Mauritius, which is the only island that they inhabited. Until recently, the dodo bird was the symbol of extinction, but that may not be for long. In 2013, Stewart Brand proposed at his TEDtalk in Washington, DC, the possibility of bringing back the dodo, among other species that are now extinct. How will this be accomplished? They plan to splice collected DNA from dodo specimens with that of their closest living relative, the Nicobar pigeon. However, at this point, like many bird species that they want to bring back, it is more of a question of whether the surrogate mother bird will tend to eggs or young, as they differ from her natural offspring.

Tasmanian Tiger

One of the current predators being looked at to bring back from extinction is the Tasmanian tiger. Last seen in captivity in the 1930s, the Tasmanian tiger, also called the thylacine, went extinct in its native Australia due to overhunting by farmers protecting their sheep (Strauss, 2019). Interestingly, the thylacine was the largest predatory marsupial in all of Australia. Recently, the Lazarus Project, overseen by Michael Archer, has been striving to "de-extinct" many animals, including the Tasmanian tiger ("Bringing Australian Animals Back To Life," n.d.). Unfortunately for the scientists with the Lazarus Project, they have only been able to recover some of the genes from the Tasmanian tiger, so the most they can hope to achieve is a Tasmanian tiger hybrid. However, they may not even need to 'de-extinct' the Tasmanian tiger. And that is because there is reason to believe that there may still be some out in the wild. Various pieces of evidence, from droppings to videos, of animals of astounding physical resemblance have surfaced from credible sources. Check out the video listed in the appendix and tell us what you think (Milman, 2013).

Quagga

This is a species that is currently undergoing de-extinction that I have been following for over 15 years. A subspecies of the zebra native to South Africa, the quagga went extinct in the wild in 1878, with the last individual dying in captivity in 1883. Hunted for their unique and highly coveted pelt, as well as by farmers who viewed them as a nuisance, quaggas were quickly no match and were quickly hunted out of existence. Being a colour variant of the Plains Zebra has allowed for successful efforts of selective breeding to create a new subspecies of the quaggas. Most notably by the Quagga Project. Started in 1987, the Quagga Project, armed with the knowledge that the quaggas were a subspecies of the Plains Zebra (rather than a separate species altogether), made "rebreeding theoretically possible".

> *No splicing required....so long as the quaggas original traits stripelessness, background colour, still lurk[ed] hidden in the Plains Zebra DNA variation, we think we can do it.*

And as of July 2016, the project has produced 10 Rau Quaggas through their breeding program, and established 12 different breeding groups that can be seen on private lands and reintroduced to three nature reserves. After following this project for so long, I think I feel just as excited as the volunteers of the Quagga Project every time they post about one of their new foals on their Facebook page.

Bucardo

This is one project that has made people question whether we should be using cloning and gene splicing to recreate extinct species. Recently extinct, the last individual of the bucardo species, otherwise known as the Pyrenean Ibex, was a female by the name of Celia, who passed away in January 2000 when a falling tree branch hit her (Choi, 2009). Before her death, they had extracted some samples of her skin cells (Choi, 2009). They then removed the nuclei from the cells (put simply, this is the part of the cell that contains the DNA), and were able to embed them into egg cells taken from goats (a close living relative of the bucardo) with all the goat DNA removed. Around 208 bucardo embryos were produced, which were implanted in surrogate goat mothers (Choi, 2009). Only 7 became pregnant, and only 1 gave birth to a live bucardo clone. It lived for 10 minutes, during which time it struggled to breathe. After its death, a dissection was performed, and it was discovered that it had a faulty lung due to a genetic defect, which is common in clones (Choi, 2009). And for a second time, the bucardo disappeared from the face of the Earth.

As advancements in genetics and cloning move forward, we have to remember to ask more than "Can we accomplish it?" and take a page out of Dr. Malcolm's book: "[We are] so preoccupied with whether or not [we] could, [we] didn't stop to think if [we] should." And with this, we'll have to look into how reintroducing these species will impact the ecosystem and those that are still in it. Animals such as the quagga already have to be controlled and moved around to avoid overgrazing in nature reserves. This will be the case for many grazers that we want to bring back. As for the predators, we will have to look into how it will impact those that used to be and could become their prey. Without asking these questions, we could easily make the Jurassic Park franchise a reality. Sometimes you have to live with the guilt of past actions because trying to fix them could only make things worse. And remember: Keep Looking for the Science in Things.

<u>**Footnote:**</u>

Since I last wrote this blog post, scientists have continued their work on making various extinct species "de-extinct". But here's an update on the projects we looked at back in 2018:

- One biotech company, Colossal Biosciences, has picked up the mantle of recreating the woolly mammoth. They started work on their current project in 2021, and after securing additional funding to the tune of $60 million in 2022, they stated in a press interview that they hope to have the woolly mammoth return by 2027. From there, once they have a cold-resistant animal, they plan on releasing it to its former habitat, Siberia, in an effort to help preserve the ecological health of the Arctic region.

- Colossal Biosciences is at it again. On January 31, 2023, Colossal Biosciences announced that they had begun work on resurrecting the dodo with the launch of their new Avian Genomics Group. This new branch of Colossal comes as a result of a whopping $150 million investment. The team is currently working on perfecting the cloning methodology, as Dr. Beth Shapiro and her team managed to wrap up a complete reconstruction of the dodo's genome in 2022. At present, there is no anticipated return date for the new dodo.

- Honestly, when I set out to write this footnote, I didn't anticipate it being mostly about Colossal Biosciences and their projects, but here we are (I should have labelled this section "The 'Colossal' Update to the De-Extinction Project"). This time with bringing back the Tasmanian tiger, otherwise known as the Thylacine. After their announcement in September 2021 of their plans to bring back the woolly mammoth, Colossal announced their plans to collaborate with Andrew Pask, an Australian scientist who is the head of the University of Melbourne's Thylacine Integrated Genetic Restoration Research (or TIGGR) Lab, in order to bring the Tasmanian tiger back from the dead. At the present, scientists are currently working on attaining a complete genetic sequence, and from there, they hope to move on to bioengineering a new Tasmanian tiger-like creature, with hopes of it being reintroduced to the Thylacine's former habitat.

- As of August 2020, the Rau Quagga that are a part of the Quagga Project are now distributed amongst six regional herds: Elandsberg Nature Reserve, Pampoenvlei, Vlakkenhuiwel, Nuewjaars Wetland SMA, Arc en Ciel, and Middelpos. Since 2018, 5 foals have been born to the Rau Quagga herd that calls Elandsberg Nature Reserve home, and 1 foal has been born to the Rau Quagga herd that calls the area around Pampoenvlei home. In just 2020, four foals were born at Arc en Ciel. The project now has a registry and studbook as herds are being expanded beyond the Quagga Project itself.

- After the failure of cloning the Bucardo, various projects have been undertaken to reintroduce ibex, any ibex, back to the Pyrenees in order to stabilize the health of the ecosystem. Because when a species is removed from a certain ecosystem, it can result in the collapse or negative shift of that ecosystem, as was the case with the removal of wolves from Yellowstone Park. In 2014, scientists managed to successfully introduce a subspecies of ibex to the French Pyrenees (Searle, 2021). Again in 2019, scientists introduced Iberian ibex to the French Pyrenees outside the town of Accous. There is no news of reattempting to clone the Bucardo; however, scientists monitoring the French Pyrenees populations have stated that with each generation, the ibex are looking more and more like the Bucardo.

Dissecting Frankenstein's Monster

(Originally Published October 30, 2018)

Halloween is just around the corner, and with that, everyone has begun celebrating their favourite ghouls and goblins, whether it's dressing up like them or binge-watching their favourite horror movies. Modern horror movies, otherwise known as psychological thrillers, can be quite easily explained, with or without science. Whether the killer is psychologically disturbed or seeking revenge, we can quickly conclude how they came to be the way they are. So today's post isn't about the science behind these horror films, but rather one of the silver screen classics. Today's post will be about the science behind Frankenstein's Monster.

Clearing Up Some Facts

Before I start dissecting Frankenstein's Monster, I feel the need to clear up a common misconception among people who haven't read the Mary Shelley classic (if you haven't read the book, do it; it is one of my favourite novels of all time) or watched any of the original black and white films: Frankenstein is not the monster made up of dead human body parts. Frankenstein is the doctor who made the monster. Ok, now that we have that cleared up, let's discuss this monster.

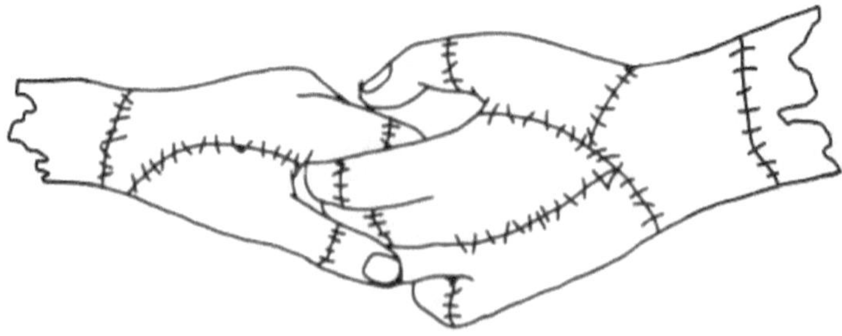

Making A Monster

In the words of Mary Shelley, Frankenstein's monster is comprised of body parts from various corpses stitched together by Dr. Frankenstein himself. Upon completing this jigsaw puzzle of a project (which he leaves rather vague to avoid others repeating his mistake), Frankenstein then brings his creature to life through a significant jolt of energy provided by lightning (as seen in the original black and white films). But is it possible? Could you assemble a body from "spare parts" and bring it to life? No, but also maybe.

Reversing Death

Victor Frankenstein selected his "pieces" from dead bodies from graves as well as medical schools (depending on which version of the story you take). These bodies have been dead for some time, on a scale of days, and then stored for many months, and the human body is declared irreversibly dead within 10 minutes of the blood flow stopping (Scott, 2013). However, there are exceptions. There are people who have experienced hypothermia, had their hearts stop beating, and been revived after more than 45 minutes of being deemed dead (Mohney, 2016). This hypothermia principle has been used to start the field of cryogenics, which allows people to freeze their bodies (or heads) so that they can be revived when the cure for their illness has been developed. Yet, this is not the case for Frankenstein's monster, and those who contributed to the body were very much dead.

Matching Parts

Then there is the problem of compatibility. Each part (or parts) is from different people, which modern medicine has taught us could have different blood and tissue types. It is essential to know the blood and tissue types of the donors (which weren't widely considered or even studied at the time of Frankenstein) to avoid rejection of the donor part. That is why many people have great difficulty finding a donor when they need an organ transplant because they are not a blood or tissue type match ("Tests you need before surgery," 2009). Even when there is a match, the body can still reject the donated part as it views this introduced component as foreign. This is why those who receive transplants also must take anti-rejection medication for the remainder of their lives to prevent the body from attacking that transplanted part ("Immunosuppressants," n.d.). So with different parts and no immunosuppressants, there is a very good chance that the body would not be able to be revived or would quickly die after being taken off the energy source, as the body's immune system would attack itself. It would not be able to live to the length that it did in the story.

Blood Typing is No Joke

While we are on the topic of compatibility, we need to talk about blood. As we previously discussed, blood typing was not discovered or studied until the early 1900s, almost 100 years after Frankenstein was written and 200 years after Frankenstein was set ("The history of blood types," n.d.). So there is the issue of incompatible blood (as well as other fluids) being introduced into the monster. And when an incompatible blood type is introduced into the body, there can be massive problems. These include extensive blood clotting (which can lead to blocked vessels, heart attacks, and strokes), excessive bleeding, kidney problems, and potentially death (Colledge & Boskey, 2017). Now you don't need the blood to pump for it to react. As you can see in the video below, blood can solidify as soon as a reactant is introduced (in the case of the video, it is snake venom). So the incompatible blood would be jelly before they even got around to jump-start him.

The Nerve of It

Assuming Frankenstein found a way to work around all these issues, there is one final issue that needs to be considered, and that is connecting the nerves. There is a reason why those who sever their spinal cords remain paralyzed: modern medicine has yet to find a way to reconnect those vital nerves that would allow their signals, or nerve impulses, to flow once again. If modern doctors still struggle with this, the likelihood that Dr. Frankenstein accomplished this feat is highly unlikely. And depending on whether the monster's head, neck, and body were all separate pieces or one intact structure, it would also determine his ability to walk, move his arms, or even allow his heart and lungs to work. Even as of right now, in 2018, doctors are working around the World to find a way to fix severed spinal cords. There have even been some who have made claims to be able to transplant a head onto a body. These claims date back to 2015 and come from Dr. Canavero. However, there has yet to be a successful human head transplant. Many doctors even say the procedure is unethical and would result in the patient having a "fate [...] considerably worse than death" (BEC Crew, 2018).

So, even if Frankenstein managed to jump start months of dead body parts and organs, lucked out and had all the body parts and fluids be compatible, no rejection took place, and he managed to fuse the spinal cord and other nerves together, the monster would have likely died of shock on the table rather than be an intelligent being that communicated, sought out love, felt rejection, and exacted revenge, all over a span of years. That being said, Shelley's Frankenstein presents us with questions to consider as we begin to venture into the uncharted territory of medical and scientific discovery. Because we create something, do we own it? Do we, as humans, have rights over other creatures? Should we revive the dead? Is cryogenics ethical? Should we transplant heads onto bodies? Because we can, should we? Two hundred years after it was first written, this work of science fiction still poses relevant questions to the science community. Just because we can, should we? That's all from us this week. Hopefully, everyone has a safe and happy Halloween. Until next time, Keep Looking for the Science in Things.

Footnote:

At the time of writing this blog post, Dr. Canavero had a candidate who was willing to have his head transplanted onto a donor body. A gentleman by the name of Valery Spiridonov, who was diagnosed with a very rare spinal muscular atrophy disease called Werdnig-Hoffman disease. Doctors had previously given him a prognosis of roughly 5 years before the disease would start affecting his nervous system leading up to his brain. His surgery was set to happen in October 2018 (Tapalaga, 2022). However, it was later revealed in an interview on April 19, 2019, between Valery and Good Morning Britain, he opted out of the surgery because he and his wife had a child. To date, there is no evidence that a head transplant has been performed on a living human. There are rumours that Dr. Canavero has "successfully" performed a head transplant on a cadaver in China. But until now, the only evidence of head transplants has been done by researchers on mice, rats, and monkeys. In 2022, Dr. Canavero had an article published in Surgical Neurology International discussing the feasibility of performing a whole brain transplant and how he would go about carrying out the procedure. So it seems that the doctor may have shifted his focus to new endeavours. At least for now.

Space

Nibiru: Destroyer of Earth

(Originally Published April 23, 2018)

First of all, I want to congratulate everyone for continuing to exist today. As silly as that sounds, there is a group of people who believe that the World is going to end today, Monday, April 23, 2018. And Nibiru is to blame.

What is this Conspiracy All About?

Ok, who or what is Nibiru, and how is it supposed to end the World? Nibiru, also known as Planet X, is a giant, rogue planetoid that many believe will collide with or have a near-miss with Earth that will cause a cataclysmic pole shift ("Nibiru cataclysm," n.d.). Believe it or not, April 23, 2018, is not the first time that Nibiru has been expected to spark the end of days. In fact, it was originally mentioned in the works of Zecharia Sitchin, from his studies of ancient Sumerian and Babylonian mythology ("Interview with Zecharia Sitchin," n.d.). However, he never claimed that Nibiru would collide with or disrupt Earth, but rather that it could be a planetoid that possessed an alien species that had interacted with humans in 556 BC and were expected to contact us again in 2900 AD after Nibiru had completed its 3,600-year orbit (Russell, 2002). Whether you choose to believe ancient Sumerian and Babylonian mythology or not is up to you (I guess we'll have to wait until 2900 AD, according to Sitchin, before we can know for sure), but if Sitchin did not propose a cataclysmic end of the World, where did this idea come from?

How did the Conspiracy Start?

The Nibiru apocalypse was first mentioned by Nancy Lieder, the founder of Zeta Talk, in the late 1990s. Lieder is a self-claimed contact of the extraterrestrial civilization Zeta Reticuli, who communicate with her via an implant in her brain. It was through this means of communication that Nancy Lieder was alerted to the dire situation of

Nibiru being on track to disrupt our planet (Wolchover, 2011). However, the original date of impact/disruption was not April 23, 2018. It's been May 2003, and December 21, 2012, and more recently, September 2017, and October 2017, and November 2017, and finally April 23, 2018. And this idea has since spawned a life of its own along the way, from the Hale-Bopp comet being a cover-up for it to the formation of cults, such as Pana Wave Laboratory, being the suspected occurrence at the end of the Mayan calendar on December 21, 2012, to finally, being taken over by David Meade, a conspiracy theorist and self-proclaimed "Christian numerologist." But what should we even believe? We're going to work through both the evidence supporting and disproving this notion, and then we can decide for ourselves rather than just clicking "share" on our social media.

Evidence Supporting the Conspiracy

Recent evidence in support of Nibiru and its imminent collision course comes from David Meade, who states Revelation 12:1-2 from the Bible as his evidence (Dangerfield, 2018). Now, whether you believe in Christianity or any religion at all is beside the point. We are not questioning your beliefs or the validity of religious texts. Let me show you. Part of the passage that David Meade calls forward as evidence is a "great sign appearing in heaven". That is extremely vague; it could be Nibiru, or it could be an angel, or God himself, or an eclipse, or a weird-looking cloud, or the sky changing a funny colour, or skywriting, etc. So to make such an extensive claim that that passage gives definite proof of Nibiru doesn't really hold water, as the saying goes. Also, the second piece of evidence that we have been given is from Nancy Lieder, who has stated that an alien race has informed her of this event through a brain implant ("ZetaTalk: Communications," 1995). Well, we don't have any MRIs or brain scans of Nancy, so we can't say if she even has a brain implant. But I think the most notable thing to take away from her message is that she has incorrectly identified the date more than once. Finally, there are pictures. Wait, what? There are pictures that have surfaced on the internet that show Nibiru next to the Sun. Now, before we start panicking, let's look at the evidence to see if this is all just an elaborate hoax or not.

Evidence Against the Conspiracy

If the flaky evidence supporting the Nibiru Event isn't enough to at least make you skeptical, let's look at the evidence provided by NASA to reassure us that we don't need to fear the end of the World, just yet. First of all, if Nibiru, or Planet X, as it is being called by some, was on a collision course with Earth, we would have been able to see it with the naked eye by now, and NASA would have been monitoring it for over a decade by now (Solar System Exploration Research Virtual Institute, 2011). This observation would be possible as Nibiru would have effects on the orbits of the outer planets in our solar system. How do we know this? Well, we, in fact, have a ninth unobserved planet in our solar system (which is not Nibiru) ("Hypothetical Planet X," n.d.). We know it's there because it interacts with the outer planets to the point where it actually affects their orbits; we just haven't seen the actual planet responsible for these actions yet. Finally, those photos I previously mentioned - yeah, they are of a phenomenon called lens flare (Sutherland, 2018). Lens flare happens when you shine a bright light on a camera and it creates a reflection within its series of lenses, and it appears like an orb or series of orbs in your picture. This can even be accomplished using your phone's camera (if you do try this at home, please, Please, PLEASE, wear proper eye protection and do not stare directly at the Sun). You may have even seen a very similar looking natural occurrence called sundogs, when the Sun reflects, sometimes off the snow, and creates multiple "Suns" in the sky in a halo (Fienberg, 2006).

Conclusion

So do we need to worry about Nibiru or Planet X colliding with Earth at some point today (April 23, 2018)? Based on the evidence, or lack thereof in some cases, I wouldn't be too concerned about it. But this provides a perfect opportunity to look at what we believe on social media and even what we share with others. So when in doubt, check it out. What evidence is there? Who are the sources of these pieces of evidence? Can you trust them, and if so, how? That's all for today. Until next time, Keep Looking for the Science in Things.

<u>Footnote:</u>

Well, clearly, the world didn't end on April 23, 2018. However, Nancy Lieder and her Zeta Talk followers are far from putting their end-of-the-world narrative to bed. A quick search of a ZetaTalk supporters Facebook page shows that not only do they continue to talk about Nibiru, but they also talk of other cataclysmic events, such as the "Jolt," "Disease X," which is caused by the red dust being shed by Nibiru, the increase in "Apocalyptic sounds," various financial collapses, and cataclysmic events as a result of the pole shift, to name a few. In 2020, they announced an upcoming global blackout in 2022 that would be caused by Nibiru's position in relation to Earth. I don't know about you, but I don't recall there being a global blackout, even a momentary one. So that's another non-event from the ZetaTalkers. As of December 15, 2021, the ZetaTalkers have announced that the poles will shift in 2026, which would spell the end of days. My personal favourite quote from their announcement is "The Pointy Heads in Antarctica who facilitated the Moloch child sacrifice ritual have left the Earth."

So should we be worried about the pole shift happening in 2026? According to a National Geographic article published in February 2018, they are currently monitoring various occurrences related to Earth's poles: The North Magnetic Pole is going through the Northern Hemisphere at 55 km per year to the northwest, and part of the magnetic field has reversed in the Southern Hemisphere. At present, scientists do not have enough information to form a conclusion on when we can expect a pole shift; however, when they are able to receive a timeframe, I do not see them withholding that information from the general public. In addition, ZetaTalk is one of the last places that I would rely on their predictions due to the accuracy of all their past predictions. So my plan for 2026 is to stock up on tin foil, as it seems to be a hot commodity these days

A Co-operative Mission of Firsts: InSight & Mars Cube One

(Originally Published May 2, 2018)

This Saturday, May 5, 2018, will be the first successful Mars landing mission since 2011, the InSight and Mars Cube One ("InSight Launch Press Kit," n.d.). NASA and JPL (Jet Propulsion Lab) have banded together to launch this two-pronged mission. Not only is this a massively cooperative mission, but it is also a mission with many exciting firsts in understanding Mars and other planets as well.

This mission will be crucial in helping scientists understand the formation and early evolution of all rocky planets. This includes Earth, Venus, and the Moon. This is the first interplanetary launch to take place away from Cape Canaveral, with the launch taking place at Space Launch Complex 3, at Vandenberg Air Force Base in California ("InSight Launch Press Kit," n.d.). Both InSight and Mars Cube One will launch from Vandenberg on the Atlas V 401 rocket, but they will separate after launch and embark on Mars separately, with an expected arrival date of November 26, 2018. So it only seems appropriate to break down the discussion of the mission into two parts.

InSight

The InSight mission, in itself, will increase our knowledge base of Mars and the solar system tenfold. InSight stands for Interior Exploration using Seismic Investigations, Geodesy, and Heat Transport ("InSight Launch Press Kit," n.d.). This is the first mission conducted to study Mars' deep interior. During the course of the mission, InSight will study the seismology, heat flow, and radio science of Mars. In other words, scientists will be taking the "vital signs" of Mars for the first time since the planet was formed 4.5 billion years ago. It will be the first time that a magnetometer will be used on the surface of Mars, and InSight will be the first spacecraft to use a robotic arm to grasp instruments on another planet ("InSight Launch Press Kit," n.d.). It is scientists' hope that by studying the interior of Mars, we will discover additional information on Earth's and Moon's interiors and how they were formed. And from this, it could assist in our search for planets outside of our solar system that could support life.

Another first that InSight is hoping to accomplish is detecting and measuring "marsquakes," or the Mars' equivalent of earthquakes, which hasn't been attempted since the 1975 Viking mission ("InSight Launch Press Kit," n.d.). However, unlike earthquakes, marsquakes are speculated to be caused primarily by volcanic activity and cracks in the planet's crust, whereas earthquakes are due primarily to tectonic plate movement. By studying these marsquakes, we will be given a look into how the waves travel through the layers of the planet, how deep those layers are, and what they are made of.

Another significant piece of information that we are hoping to get is how the magnificent Martian volcanoes are formed, including the volcanic plateau of Tharsis. InSight will utilize a self-hammering heat probe that can burrow up to 5 meters (previous missions only reached an average depth of 3 meters) into the soil to measure the heat flow of Mars for the first time ("InSight Launch Press Kit," n.d.). This data will allow us to see how the internal energy of the planet drives the changes on the planet's surface.

Mars Cube One

Mars Cube One is the first attempt at utilizing miniaturized CubeSat technology on other planets and as a means of deep space communication ("InSight Launch Press Kit," n.d.). During the launch on Saturday, two CubeSats will be launched, with the InSight, each being the size of a briefcase. Once it successfully reaches Mars, they will relay InSight's data as they enter Mars' atmosphere and proceed to land on the surface of the planet. If successful, the Mars Cube One mission will become a new means of communication with Earth.

Stats of the Mission

Now for some cool stats on the mission:

- On May 5, 2018, the proposed day of the launch, the distance between Earth and Mars will be 121 million km.

- The height of the Atlas V 401 Rocket with payload (cargo) is 57.3 m, which is more than 2 blue whales placed end to end or as high as a 19-story building.

- Total payload (cargo) Mass: 721 kg (1590 lbs.)

- The distance that InSight and Mars Cube Ones will travel will be about 485 million km by the time they reach Mars.

- Landing date: November 26, 2018, during Mars' Winter

- Landing Site: about 4.5 degrees North latitude, 135.9 degrees East longitude, on the Elysium Planitia

- Mission Length: One Martian year plus 40 Martian days (nearly 2 Earth years), until November 24, 2020

Why Mars?

One thing you may be wondering is what is the deal with our fascination with studying Mars, especially considering we still haven't found any life on it. Well, Mars can be equated to a time capsule. Because it is smaller than Earth and its twin Venus, it has less internal energy, which results in changes occurring on the surface of the planet at a slower rate ("InSight Launch Press Kit," n.d.). So much so that Mars has remained constant, or static, for more than 3 billion years. This allows us to see the solar system's early history, which has long been lost on other planets like Venus and Earth.

Where You Can Tune In

Now that I have you excited about the scheduled launch on Saturday, May 5, 2018, where can you tune in to watch this momentous occasion? Well, there is NASA TV which is an app that you can install on your phone, iPad, and even your AppleTV. NASA JPL also has a live YouTube channel that will provide a live feed during launch. Here's the catch. The launch opportunity will begin at 4:05 a.m. PST on Saturday and will continue until the launch can occur, ending on May 8, 2018. Yes, I heard the yikes and collective sighs. You may not be a morning person, and that's ok! Because Schrodinger's Cat has you covered. For the weekend of May 4-8, we will have an event running on our Facebook page that will keep you up to date on the launch, so be sure to click "Attending" on the event to make sure you get your launch updates, as well as see the launch video, in case you miss it live.

With that being said, I'm going to leave you with some fun facts provided by NASA JPL about Mars missions and the upcoming mission. I look forward to talking to all of you at the event. Until then, Keep Looking for the Science in Things.

Fun Facts:

- Mars missions have been conducted by NASA since 1965 (involving flybys, orbits, landings, and rovings).

- Only approximately 40% of Mars missions by any agency have been successful.
    - o   The thin atmosphere makes landing difficult.
    - o   Extreme temperature swings on the surface make operations not work.

- One-way radio transit time from Mars to Earth on Nov. 26, 2018, will take 8.1 minutes as it has to travel 146 million kilometres ("InSight Launch Press Kit," n.d.).

- Expected near-surface atmosphere temperature range at the landing site during the landing will be between -100 Celsius to -20 Celsius ("InSight Launch Press Kit," n.d.).

- Total investment in the project: US investment in InSight totals $813.8 million, about $163.4 million for the launch vehicle and launch services; France and Germany invested about $180 million in InSight, primarily for the seismometer investigation and heat flow investigation; JPL and NASA invested about $18.5 million for Mars Cube One technology ("InSight Launch Press Kit," n.d.).

Footnote:

On April 24, 2022, the InSight Mars lander took its final selfie after completing its 1,211th Martian day mission. The final image taken was on December 11, 2022, and the final communication between NASA and InSight was on December 15, 2022. Already extended from its previous mission end date, InSight finally powered down as the Martian dust accumulation on its solar panels greatly hindered its ability to recharge. However, during its time on Mars, it broke ground, both literally and figuratively:

- InSight provided the clearest view of the Martian Core to date.

- InSight determined the crust of Mars was about 15 to 25 km thick and comprised of three internal layers. The top layer is about 10 km thick and less dense than the lower crust.

- InSight determined that Mars' core is about 1,800 km in radius and has a lower density as a result of lighter elements mixed with molten iron.

- InSight detected the first quake on another planet. It detected a total of 1,319 marsquakes.

- Data acquired from meteoroid impacts monitored by InSight helped scientists determine the age of Mars' surface.

- InSight helped contribute to the discovery of ice deposits on Mars as a result of a meteoroid strike in 2021.

- It was the only active seismic station beyond Earth.

- InSight had the first magnetometer on the Martian surface, and with that, it helped scientists discover that the crustal magnetization remnants of its former global magnetic field are 10 times stronger than they expected.

The end of the mission is best summed up by Bruce Banerdt of JPL: "We've thought of InSight as our friend and colleague on Mars for the past four years, so it's hard to say goodbye. But it has earned its richly deserved retirement."

MarCO-A and MarCO-B completed their mission on November 26, 2018. One was last heard from on December 29, 2018, and the other on January 4, 2018. They were the first interplanetary mission for CubeSats and show that it is possible to carry out interplanetary missions with mini-spacecraft, thus greatly reducing the costs of missions of this nature.

<u>Water on Mars?</u>

(Originally Published July 27, 2018)

After years of searching, they may have just found it. Water. Liquid water on Mars. In a study published on July 25, 2018, scientists have reason to believe that they may have finally found a significant amount of water on Mars. But how? And what does this mean for the future of human exploration of Mars?

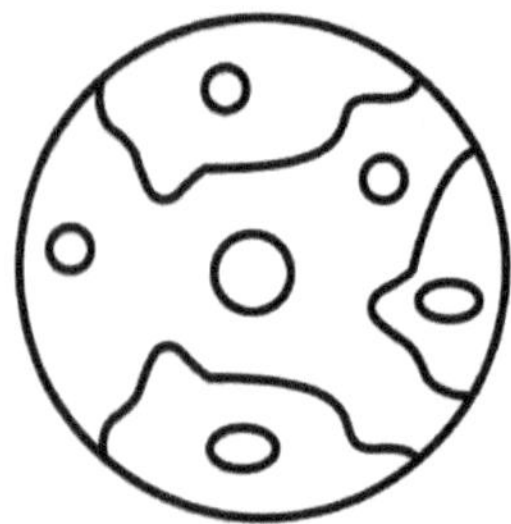

Finding Water

It is well known to scientists that Mars at one time used to be covered by oceans. But where did all the water go? Scientists are still unsure what happened to these vast bodies of water, but with this recent discovery, they may be one step closer to uncovering the truth. This is not the first time that scientists have discovered water on the red planet, but it has always been elusive; either it has been inaccessible or unpredictably on the move. Previous water discoveries include water vapour in the atmosphere, frozen solid in permafrost and polar ice caps, and occasionally seeping down crater slopes. All of which is important; it hardly makes the planet any more inhabitable. But this recent discovery has proven different.

This discovery has been a long time coming. MARSIS, the radar responsible for detecting this latest discovery, first detected shimmers back in 2008 around Mars' south pole region (Drake, 2018). This area is known by scientists for its extensive ice sheets that stack on top of one another. However, it wasn't until the radar shimmer came back that the team of scientists decided to further look into the region. The reflection was more consistent with that of Earth's oceans than our polar ice caps. It has taken over 9 years, many satellite passes over the region, and countless hours analyzing the mountains of data, and scientists have come to the conclusion that it could be one of two entities under the ice: either it's a lake or it is saturated sediment, much like an aquifer (Drake, 2018). So the final picture that scientists have painted for us is that it is "a stable reservoir of salty, liquid water," which is about 19 kilometres long and is about 1.5 kilometres under Mars' south pole (Drake, 2018).

Water In Space

Although Earth is the only planet in our solar system to have consistent and abundant water, it is not the only one to have water. Besides this recent discovery on Mars, scientists also believe one of Jupiter's moons, Europa, and one of Saturn's moons, Enceladus, could also have water.

Europa

Discovered by Galileo himself in 1610, scientists have suspected the presence of a subsurface ocean on Europa since the Voyager missions that arrived at Jupiter in 1979 (Thompson, n.d.). One of the first telltale signs that they spotted on Europa was the presence of dark bands that fit together like a jigsaw puzzle (Thompson, n.d.). They speculated that these lines had been cracks that had separated and then filled in, which would suggest the surface hadn't always been as solid as it is today. In addition to these lines, there were very few impact craters observed on Europa, considering that the planets and moons in our solar system have been constantly bombarded by meteorites over the past billion years or so (Thompson, n.d.). The lack of craters indicated to the scientists that this icy surface may be relatively recent and many of the craters may have been erased by the moon's activities (Thompson, n.d.). Then, scientists observed an unusual phenomenon. When looking at the long linear features on the moon's icy surface, they determined that they did not fit the pattern of a normal tide but rather were more consistent with an icy surface floating on top of water or slightly warmer ice (Thompson, n.d.). This led scientists to construct a model of Europa that is not frozen all the way to the deep interior.

Enceladus

During NASA's Cassini mission, scientists detected a slight wobble in Enceladus' orbit around Saturn, leading them to believe that the moon's icy surface does not extend all the way to its core ("Cassini [..]," 2015). In fact, the wobble could be due to the presence of liquid water under the ice cap. In addition, there have been findings that mist of water vapour, icy particles, and simple organic molecules are being emitted through cracks in the ice at the Enceladus' south pole ("Cassini [...]," 2015). This, paired with radar imagery, has led scientists to believe that not only is water present under the ice but that it is very abundant, more or less like a global ocean under the ice.

The Big Deal About Water

Water in space is a rare commodity. So the presence of it in any significant quantity gets astronomers and scientists excited because it can mean a multitude of things.

First of all, we may not be alone. I'm not saying that there are merpeople on Europa or aliens on Mars that vaguely look like E.T. But rather bacteria and microorganisms that have found a way to survive in some of the harshest conditions known to humans. And we know that it is possible. Tardigrades, or water bears, live in some of the most impossible conditions on Earth. They can be found on moss sandwiched between two slabs of sidewalk (among other locations), and we have subjected them to extreme pressure and even launched them into space to see how

they would fare against solar radiation. Unaffected. Then there
are archaebacteria (Baxamusa, 2018). These primitive bacteria can
withstand some of the harshest conditions on Earth, from living in
conditions such as being in environments completely without oxygen to
living in highly salty waters, such as those in the Dead Sea, to thriving in
hot and highly acidic conditions, like in hot springs and inside geysers.
If we have organisms like this, what's to say there aren't any on, or were
on, Mars, Europa, or Enceladus?

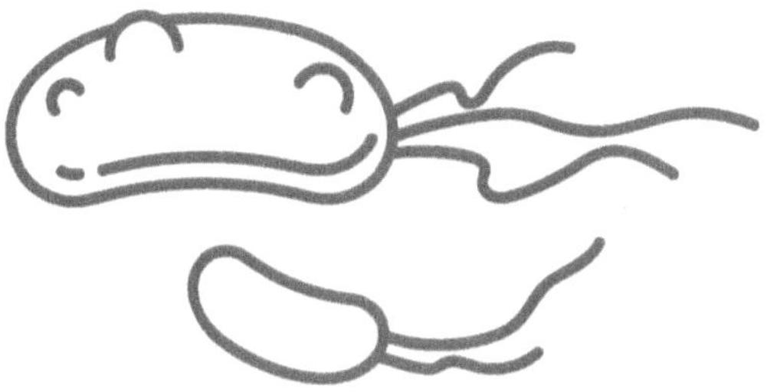

Secondly, even if there is no life there now, who's to say it can't support
life in the future? As the Earth faces the issues of climate change,
overpopulation, and food shortages now and in the future, many people
are looking to other planets to become Earth 2.0. Even if the abundant
supply of water is not freshwater, we have developed and continue to
develop technology to purify and desalinate water. This water could
then be used to supply colonists with usable water for drinking, bathing,
agriculture, and many other applications.

This recent discovery, although not unlike other discoveries, provides
new hope for researchers and people in general as we move forward in
humankind's ventures into space. It may prove that we are not alone or
that we could live on planets other than Earth. Regardless of where you
stand with the implications of this discovery, we can all agree that it is
significant, and those who have spent years pouring over the data should
celebrate this achievement. That is all from us today. And Remember to
Keep Looking for the Science in Things.

<u>**Footnote:**</u>

Since the discovery of ice on Mars in 2018, when this blog post was written, scientists have determined several more locations on Mars, which include both poles and various craters and gullies. Between May 2018 and February 2021, a large deposit of hydrogen was discovered in the Valles Marineris canyon system (a canyon system that dwarfs the Earth's Grand Canyon) by the ExoMars Trace Gas Orbiter (Strickland, 2021). To make matters even more interesting, this region of the planet reaches temperatures that make it difficult for ice to remain frozen. Between May 2021 and May 2022, China's Zhurong rover returned data that suggests that Mars had liquid water as recently as 400,000 years ago, based on their determination that the formation of the ridges and fissures on Utopia Planitia was likely due to the thawing of snow or frost (for context, homo sapiens evolved roughly 300,000 years ago). In addition to this revelation, Zhurong has sent back data suggesting that Mars' former microbe population likely had to be adapted to very salty environments. Pair this data with the data that was unearthed with the assistance of InSight, and we are one step closer to allowing for sustainable habitation on the Red Planet.

Shooting For The Star: Looking at the Parker Solar Probe

(Originally Published August 7, 2018)

In the early morning hours of August 11, 2018, a United Launch Delta IV-Heavy is expected to launch from Cape Canaveral Air Force Station, carrying a space probe that is the first of its kind and is more than 50 years in the making. The Parker Solar Probe. Scientists hope that with this launch, we will gain a better understanding of the star that powers our solar system, our Sun. Today's post will give you the lowdown on the Parker Solar Probe mission, what we hope to learn, and why it's important.

Mission Objective

Through the launch of the Parker Solar Probe, we hope to collect new data about the Sun's activity and use it in forecast models of major space weather events that could affect us here on Earth. NASA and those involved in the mission (John Hopkins University Applied Physics Laboratory and the Parker Solar Probe Instrument Principal Investigators) plan to gain this information by following the flow of energy from the Sun to better understand how the area immediately surrounding the Sun, the solar corona, heats up and determine what causes solar winds to accelerate ("Parker Solar Probe," n.d.). In addition, they hope to determine the structure and dynamics of the plasma and magnetic fields, which are the root cause of solar winds, and look into what causes the acceleration and transportation of the energetic particles emitted from the Sun.

Journey to the Sun

Following its launch this Saturday, the Parker Solar Probe will journey towards the Sun using a new technique. It will perform 24 orbits (each lasting around 88 days) around the Sun with each orbit becoming smaller and smaller, pulling the probe within 6.16 million kilometres of the Sun, which is seven times closer than the current holder of the closest solar pass, which was achieved by the Helios spacecraft ("Parker Solar Probe," n.d.). In order to achieve this, the probe will travel for seven years, using Venus' gravity (seven times) to "slingshot" through its orbits ("Parker Solar Probe," n.d.). The first Venus flyby is expected to occur on October 2, 2018, at 23:45 UTC, and achieve the first perihelion (the closest point to the Sun in an orbit) on November 5, 2018, at 18:33 UTC ("Parker Solar Probe," n.d.). In the final orbits, the Solar Parker Probe will soar past the Sun at around 692,018 km/h and endure temperatures around 1,400°C (and yet, it will manage to keep its measurement instruments at room temperature or around 20°C) ("Parker Solar Probe," n.d.).

Behind the Name

The Parker Solar Probe received its name from physicist Eugene Parker, who in the 1950s proposed concepts on how stars give off energy ("Eugene Newman Parker," n.d.). He was the first to identify solar winds and explain how plasmas, magnetic fields, and energetic particles make up these winds. Also, he has speculated on how the solar corona heats up. 60 years later, this mission will help shed light on these theories. With Dr. Parker able to oversee these achievements.

Why the Sun?

So why is studying the Sun important? As we all know, much of what makes up life on Earth is driven by the Sun: weather, ocean currents, crops, energy, and life. Without the Sun and how it works, life as we know it would not be the same. However, the Sun can be unpredictable at times. Solar flares are a prime example of how the Sun can suddenly unleash its fiery potential. From causing a spike in Earth's temperature and altering the shape and chemistry of the atmosphere to releasing bursts of ultraviolet radiation that can cause an increase in Earth's ozone and interfere with our electromagnetic frequencies (e.g. cell phone, computer, radio signals), by understanding how the Sun "works," we can help to predict and react to solar flares and lessen their impact on Earth and our lives ("Solar storms," 2017).

This upcoming Parker Solar Probe mission can help us better understand the heart of our solar system, the Sun. And through this better understanding, we can better react to anything the Sun may throw our way. On top of this, it could also prove to be revolutionary in our ability to travel closer to stars and in how space missions could use gravity assistance and decreasing orbits to travel. The next seven years will prove to be quite insightful as we venture into unknown territory. That's all from us today. In the meantime, Keep Looking for the Science in Thing

<u>**Footnote:**</u>

As of July 2023, the Parker Solar Probe is on its 16th orbit of the sun. It is currently travelling at around 70,293 km/h and is roughly 103,849,867 km from the sun (based on mission performance modelling and simulations). I will include a link in the appendix so you can see real-time data regarding the Parker Solar Probe's position and speed, as well as all of its discoveries. However, at the time of writing this footnote, here are some of PSP's discoveries since 2018:

- The PSP has provided valuable data on the asteroid Phaethon and the Geminids meteor shower. This is the first time that the Geminids have been studied anywhere other than on Earth, and scientists now believe based on the data provided by the Parker Solar Probe that the Geminids were created as a result of Phaethon experiencing a "violent, catastrophic event - such as a high-speed collision with another body or a gaseous explosion." (O'Regan, 2023)

- The PSP has allowed scientists to study how solar winds are generated through a process called magnetic reconnection and even detect "details that are lost as the wind exits the corona" (Sanders, 2023). These details include PSP detecting "streams of high-energy particles that match the [...] flows within coronal holes," regions where scientists now believe fast solar winds originate (Sanders, 2023). When studied from Earth, the solar winds only appear as a chaotic yet "homogeneous"flow of magnetic fields and charged particles, which gives very minimal details on how they are actually formed.

- The Parker Solar Probe has broken the record for the fastest human-made object (travelling at about 586,863 km/h) and being the closest to the sun at 8.5 million km away. However, the PSP will beat all of those records again when it's on its final orbits around the sun, reaching an estimated 692,018 km/h and being only 6.3 million km away.

Sadly, Dr. Eugene Parker passed away on March 15, 2022, but his name and his work regarding solar winds will live on as the Parker Solar Probe is set to continue its mission orbiting the sun until its final orbit on June 19, 2025.

The Ultimate Guide to the Perseid Meteor Shower 2018

(Originally Published August 9, 2018)

The launch of the Parker Solar Probe is not the only astronomical event that is happening this weekend. On the nights of August 11 to 12 and 12 to 13, the Perseid Meteor Shower will be lighting up the night sky. In today's post, we will be discussing all you need to know about the Perseid Meteor Shower and where and how you can best watch the meteor shower for yourself.

What is the Perseid Meteor Shower?

The Perseid Meteor Shower occurs every Summer as the Earth passes through the tail of the Swift-Tuttle Comet. Discovered in 1862 by Lewis Swift and Horace Tuttle, the Swift-Tuttle comet is made up of dust, ice, rock, and dark organic matter. And it is the dust and debris from the tail of the comet burning up as it enters Earth's atmosphere that creates the Perseid Meteor Shower. The core of the Swift-Tuttle Comet is 26 kilometres wide and passes by the Earth every 133 years as it orbits around the sun (Lewin, 2016). The last time it passed the Earth was in 1992. Prior to this passage, scientists believed that it was due to collide with the Earth in 2126; however, thanks to measurements taken in 1992, we have no reason to fear. It will not be colliding with the Earth anytime soon.

Where to Best View the Meteor Shower

This year, the Earth will be passing through the Swift-Tuttle tail between July 17 and August 24, reaching its densest around August 12, where around 50 meteors an hour can be seen. The best locations to view the meteor shower include the entire Northern Hemisphere to -35 degrees latitude in the Southern Hemisphere (the temperate latitudes). However, one thing to bear in mind this Summer when viewing the meteor shower is the high level of smoke in the atmosphere from the fires in Western North America. So visibility in the immediate area of the fires as well as those areas downwind of the fires may be unfortunately limited (Lada, 2018).

Meteor Shower Viewing Tips

For those that are wishing to view the Perseid Meteor Shower, there are some ways to make your viewing experience enjoyable:

- Best Time for Viewing: The best viewing time to see the meteors is after midnight in the Northern Hemisphere and just before dawn along the northern horizon in the Southern Hemisphere.

- Stay Away from the Light: To best see the meteor shower, head out of town. Any light from buildings, street lights, or other light pollution can hinder your ability to see the meteor shower clearly.

- Get Comfy: It can take around 20 minutes for your eyes to adjust to the dark, so get comfy. Bring blankets to lay on while you are gazing at the stars.

- Watch with Others: Get your friends and family together to watch the meteor shower together. More eyes on the sky means you are less likely to miss out on meteors.

This weekend, Schrodinger's Cat will be scanning the skies for the Perseid Meteor Shower, and we hope that our readers will join us. Remember to Keep Looking for Science in Things.

Technology

Making The World More Accessible, One Piece of Tech At A Time

(Originally Published May 16, 2018)

As you may or may not know, tomorrow, May 17th, marks World Accessibility Day, a day that highlights the importance of providing and developing accessibility for those who need it. Be it in innovative technologies, further research and knowledge, or reforms in restricting legislation. Approximately 10% of the World's population (around 650 million people) live with a disability (Disabled World, n.d.). That means the average person has at least one person in their life who has some level of disability, be it a visible one or an invisible one. Throughout human history, there have been key innovations to improve accessibility and independence for those who would otherwise struggle. Such as mobility aids (wheelchairs, crutches, prosthetics, etc.), communicative aids (braille, sign language, text-to-speech, hearing aids, etc.), learning aids, and much more. In today's post, we are highlighting some accessibility technologies that are just in the fledgeling or prototype phase, but show much promise to improve accessibility in society.

1. GyroGlove by GyroGear

The first piece of technology is still in the prototype phase but shows a great deal of promise for those who have uncontrollable hand tremors, like those with Parkinson's disease. GyroGloves use gyroscopes to counteract the tremors' movements and allow the individual to regain the use of their hands. Allowing the individual to feed themselves, write, draw, and perform other fine motor skills. But how do gyroscopes even work? Well, a gyroscope is a spinning mass (a disc or wheel) that spins about an axis that can freely move (McGlashen, 2015). Because of this, they maintain their orientation (positioning) regardless of a force (such as gravity or a tremor) acting against them, trying to knock them off balance. With the gyroscopes attached to the gloves that the individual wears, they help to steady their hands against the tremors, counteracting the uncontrollable force, and allowing them to use their hands.

2. The Snail by Wonkook Lee

The next piece of technology was developed in order to improve upon the innovation of Braille. The Snail Braille Reader is an electronic voice converter that the user rolls over Braille, and it converts the bumps into speech, that the user can hear out loud or through a Bluetooth earpiece (Seth, 2011). How it accomplishes this is through a pressure sensor wheel that feels each individual bump and translates it into spoken word. And to keep the reader from straying from the line of braille, it has a guide wheel on each side of the pressure wheel to help keep it centred on the line being read. The innovators behind this technology claim that their technology speeds up not only the individual's ability to read the text but their ability to learn; as learning to read braille is learning a new language. With their device, the user can be back to reading their favourite book, or textbook, within an afternoon and at a rate similar to a sighted reader.

3. My Buddy Tag Lockable Tracker

You may have seen many similar pieces of technology on the market, but there are a few key features that I found that make the My Buddy Tag Lockable Tracker stand out. For those of you not familiar, My Buddy Tag Tracker is a wristband-style tracker that allows family members and caregivers the ability to monitor individuals that are a flight risk or get lost easily, such as those with Autism, Dementia, or Alzheimer's, and even small children. The family member or caregiver installs the app on their phone and sets a proximity (how far the individual can go away from you before an alert is issued). If the individual wearing the wristband leaves the proximity, an alert is sent to the caregiver's cell phone, and they can track them on the map while the individual is within Bluetooth range of the cell phone. Once the wearer leaves the Bluetooth range an email alert of the last seen location and time is sent to the caregiver. In addition to the proximity feature, there is a notification if the wearer falls in water, as the tracker senses the water and relays the alert to the cell phone, reducing the risk of accidental drowning. As well, the tracker has a panic button that the wearer can activate should they become disoriented or run into trouble, and the caregiver receives an alert with the last seen location and time. Many of the tracking solutions that I have seen on the market have one noticeable flaw. They are easily removed. This tracker has a tamper-proof screw-type clasp that requires a screwdriver to remove. Also, for those who have sensory issues or sensitive skin, the tracker can be removed from the wristband and placed in a pocket, on a necklace, or threaded through shoelaces. The My Buddy Tag Tracker is definitely one piece of technology to keep an eye on as it develops further.

4. Cochlear Implants

Not necessarily a very new innovation, but even I will admit that prior to conducting research for this post, I didn't even know how cochlear implants worked ("Understanding Cochlear Implants," 2016). Here's what I learned: Normally, sound waves enter an individual's ear and hit the eardrum. This, in turn, causes the tiny bones in the inner ear to move, which moves the fluid in the inner ear. This fluid transfers the motion to tiny hairs in your ear called cilia, which create electrical signals to the brain through the hearing nerve. Now, when someone sustains inner ear damage, this process can be disrupted, and the sound wave is not converted into a signal that the brain can interpret. Traditionally, hearing aids were used to assist those with hearing impairments; however, they only amplify the sound. For some cases of hearing loss, making the sounds louder doesn't help. That's where cochlear implants are a game-changer. Much like with many hearing aids, there is a piece that sits behind the individual's ear. In cochlear implants, this is a sound processor, which captures sounds and converts them into a digital signal. This signal is then transferred to the internal implant, which converts the signal into an electrical pulse. This electrical pulse is sent to an electrode array, which is surgically implanted in the cochlea (inner ear). Electrodes on the electrode array stimulate the hearing nerve, which then sends the electrical signal to the brain like it should. Meanwhile, bypassing the damaged portion of the inner ear and allowing individuals to hear birds, music, and their name for the first time.

5. ReadingPen

Growing up and going through school, many of my friends had learning disabilities such as dyslexia and other disabilities that made reading and learning a herculean task (many not even having a clinical name at that point). This device could have worked wonders for them. Called the ReadingPen, this support device can be used to scan over words that the individual doesn't know or understand, and the device will say the word out loud, followed by the definition displayed on its screen. If there is a word in the definition that the user does not understand, they can look that word up as well. In addition to dictionary capabilities, it also has an active thesaurus that allows the user to look up similar and related words. Now, many of my friends had a horrible time taking tests, and it was because they would just blank on a word or their test anxiety would jumble up the letters so they could not even read it. Well, the

ReadingPen has a test mode, which turns off the dictionary and thesaurus for the purpose of tests but will still read out the word so the user can hear the word. Pretty cool, if I do say so myself.

These five innovative pieces of accessibility technology only represent a few of the inspiring concepts and ideas that are coming out every day. The only difficulty many of these projects face is the lack of funding they receive. So in honour of this World Accessibility Day, whenever you are given the opportunity to have your say in where funds are donated, please speak up for those that are also working hard to make the world a more accessible place for everyone. I hope you enjoyed today's post. In the meantime, Keep Looking For the Science in Things.

Watch Where You're Driving!: A Look At Driverless Cars

(Originally Published June 15, 2018)

We've all seen the movies where a person hops into a vehicle and it drives away on its own. All the driver, who is now merely another passenger, is required to do is sit there and enjoy the drive. Modern automobile manufacturers are claiming that they will have fully autonomous vehicles (in other words, driverless cars) on the road as soon as this year (2018), with other companies stating their models will be ready by 2020. But how do these cars even work? And what does it mean for all of us?

Companies such as Tesla, Google (Waymo), Toyota, Honda, Nissan, Lyft, Uber, and Volvo have all joined the latest innovation race. To produce the first successful autonomous and driverless vehicles. But how do they even propose accomplishing this? Each prototype is equipped with some of the latest computer software and hardware, all primed to prevent the need for human intervention while driving down the road (TheHUB, 2017). Most of the vehicles utilize a combination of LIDAR (which is a laser-based radar system), GPS, and inertial measurement unit sensors (IMU). With these systems working together, they are able to pinpoint the vehicle's position down to 0.5 cm. In addition to this system, they also have a radar control system that kicks in when an object or obstacle comes within 4.57 to 9.14 meters of the vehicle. On top of all these pieces of equipment, the car's computer "brain" has learning capabilities, which reviews camera and GPS data and allows it to predict the fastest routes. With these patterns and learning capabilities, the vehicle is able to calculate the likelihood of certain actions by the other drivers.

Now you're probably wondering, "How much will it cost for me to get my hands on one of these vehicles?" Well, according to Boston Consulting Group, the additional cost of having a vehicle capable of autopilot is an additional $5,500 USD on top of the vehicle's asking price, and to be able to be entirely driverless, you are looking at an additional $10,000 USD on top of the asking price (TheHUB, 2017). And the vehicles that possess these capabilities are not your average Toyota Camry or Honda Civic. Two vehicles that you can currently purchase with the autopilot feature are the Tesla Model S and Model X. Model S basic model with no add-ons, up to 416 km per charge max (smaller battery), you are looking at $96,650 CAD, add another $6,600 CAD ($7,900 CAD after delivery) for enhanced autopilot, and an additional $4,000 CAD ($5,300 CAD after delivery) for full self-driving capabilities. For a grand total of $107,250 CAD to have a self-driving vehicle. The Model X basic model, with no add-ons other than to make it self-driving, and a max range of 381 km on a single charge (smaller battery), you are looking at $112,900 CAD. Wow! It looks like I'll be sticking to my Volkswagen Jetta for the time being.

In addition to Tesla, Google, Toyota, Honda, Nissan, Volvo, and the others jumping on the self-automated bandwagon, the manufacturer Daimler, which you may recognize from their partnership with Mercedes-Benz, is preparing to roll out robot-controlled semi trucks (Sanchez, 2014). They hope to have them hitting the pavement as soon as 2020. These units will be fully automated, and there will be no need for a driver who has to withstand the long driving hours in all weather conditions.

Most of the current testing for these autonomous and driverless vehicles is occurring in the United States, which has made the United States National Highway Traffic Safety Administration (NHTSA) have to make some new regulations in preparation for more of these vehicles on the roadways. Even if you have no plans to own a vehicle with autopilot or self-driving capabilities (or you can't afford one), you may want to pay attention, as this will affect your driver-required vehicles as well. The NHTSA has mandated that all new vehicles must have mounted rear backup cameras, forward collision avoidance capabilities, and lane detection equipment (TheHub, 2017). They are also discussing the need for vehicle-to-vehicle communication in the future, so vehicles with autopilot and driverless capabilities can read the traffic. And we can only speculate, but as time moves on, it may become mandatory for older vehicles, including classic cars and hotrods, to have these systems installed.

Now that we know a bit about these vehicles, their price tag, and the new regulations that are being implemented, let's talk about the good, the bad, and the ugly.

First Up, The Good:

With fully automated cars, there is less human error, which can result in fewer accidents. Also, with a computer running things, the vehicles can drive at higher speeds, with less traffic congestion and better parking availability in metropolitan settings. Also, since there is no driver, just passengers, you are able to be more comfortable on long road trips and enjoy the scenery more. Because it's about the journey as well as the destination, isn't it? Driverless vehicles allow for better transportation and accessibility for those that can't drive (e.g. the elderly and those with disabilities), as you would not need a driver's license as you are not the one operating the vehicle. Some insurance analysts even speculate that vehicle insurance will become obsolete, as there will be no such thing as human error, but rather the computer will run things like a well-tuned, well, machine.

But With The Good, There is Always The Bad:

As previously discussed, the vehicle is capable of calculating the likelihood of another's actions, not outright predicting them. So accidents can still happen. Also, despite these vehicles being marketed as fully autonomous, car manufacturers are stating that the driver must stay alert and be prepared to take control of the vehicle should there be technical or sensory issues ("Autopilot," n.d.). So you don't truly get to kick back and relax on your journey. Many also speculate that there will be an increased energy need to power the additional equipment on the vehicles, so some models may emit increased levels of fuel emissions (TheHUB, 2017). We've already discussed the higher price point of not just these vehicles but also the aftermarket installation on older vehicles (whether we like it or not). And finally, it's a computer, and like all computers, it does pose the potential to get a virus or be hacked. Many are asking, "If the vehicle can be hacked, what else can the hackers gain control of?" Controlling the vehicle's GPS? Change the direction of the vehicle? Track the vehicle? Intentionally cause accidents? These are things that we have yet to find, even as real-life testing is occurring on the roads.

Now, for the downright ugly: accidents. The NHTSA claims that vehicles with autopilot and driverless capabilities have safety benefits that far exceed any cons that they may have. However, since the start of 2018, the safety of many of these vehicles has been called into question, as they have been involved in accidents that may not have occurred if there had been an attentive driver operating the vehicle. Here are some accident facts:

> **January 2018** (TomoNews US). A Tesla vehicle, while in autopilot mode, rear-ends a parked fire truck going 65 mph on a Los Angeles freeway. No one was hurt.
>
> **March 18, 2018** (ABC News). While crossing the road with her bicycle (outside of a crosswalk), a pedestrian is struck and killed by an Uber self-driving vehicle in Arizona, US. The vehicle was in self-driving mode, with a human operator sitting inside.
>
> **March 23, 2018** (ABC News). Another Tesla, while in autopilot mode, crashes into a centre barrier on a California freeway,

killing the occupant. According to Tesla, the vehicle had sent several warnings earlier in the drive to the operator to put his hands on the steering wheel, but no hands were detected on the steering wheel six seconds prior to the collision. According to the victim's family, he had complained of the vehicle moving towards the same barrier 7-10 times prior and even took the vehicle to the dealership to be fixed; however, the dealership was unable to replicate the problem. Tesla also pointed out that the barrier had been shortened not long before the incident because of a previous accident. Because of the nature of the accident, the electric battery caught fire, which fire crews spent six hours trying to put out as they could not use conventional methods.

April 2018 (KPIX CBS SF Bay Area). A few days after the Tesla accident in California, another Tesla operator submitted a video to the media, showing his vehicle moving towards the exact same barrier. This proves that this is not a flaw in a single vehicle. In the video, it was also shown that the white lines were not clearly marked on the road leading up to the barrier. This has led experts to believe that the car's sensors mistakenly think that the left lane is actually the right and will proceed to move to the left. So that means all road marking lines must be clearly visible and not obstructed (e.g. snow, debris, etc.) in order for the system to operate correctly. Most testing has been occurring in areas that would have ideal weather, such as California, Arizona, and Florida. But what happens when we move into areas of snow and ice?

And there are many other accident stories out there. Many, like these ones, may not have happened had the vehicle been operated by a person paying attention to the road and its conditions. Some car manufacturers have paused production and testing of their autonomous vehicles in light of recent events (Boudette, 2018). But with other vehicles with autopilot and self-driving abilities already hitting the road and more expected to roll out in 2020, it might be time for us and car manufacturers to take a step back and say, "Are we ready?" Is it really as safe as we want to believe? I'll leave that up to you to decide. Thank you, everyone, for reading today's post.

<u>**Footnote:**</u>

Honestly, with how intensely automotive manufacturers were pursuing driverless or self-driving vehicles, I thought in 2023 there would be a lot more of them on the road. But we are no closer to having more self-driving vehicles on the road than we were in 2018. Manufacturers have claimed "two more years, two more years." It's what we heard back in 2018, and it's what we continue to hear now, in 2023. However, I am glad that these automotive manufacturers have hit pause on the big roll-out for now, as many of the issues that we discussed in 2018 are still present in 2023. Although it wasn't completely a self-driving vehicle, I have had the chance, since 2018, to try out vehicles with lane-keeping assist, and even with that technology, I still don't 100% trust it as it relies on sensors detecting the painted lines on the roads, which here in Canada and in other parts of the world aren't always present (either due to fading or covered by snow/ice/mud/etc.) So it may be another two, five, or even 10 years before self-driving vehicles are more commonplace.

5 Canadian Innovations that Changed the World

(Originally Published July 3, 2018)

This past Sunday marked Canada's 151st birthday, which Canadians throughout the country, as well as those outside the country, celebrated in true Canadian fashion. Also on Monday, Schrodinger's Cat officially became syndicated with Science Borealis, the Canadian Science Blog Network. All this has us in a very Canadiana mood here at Schrodinger's Cat, and despite our overly polite nature, in today's post, we will be featuring some of the great Canadian innovations that we still use to this day. Sorry, not sorry.

1. Cardiac Pacemaker

Invented in 1950 by Canadian scientist Dr. John A. Hopps, the pacemaker was invented by sheer coincidence (Grant, 2017). Originally, Hopps had been working with radio frequencies to aid in the treatment of hypothermia by using the frequencies to attempt to increase the body's temperature. It was through this that he gained the knowledge that the heart could be jump-started with artificial means. At the time, the pacemaker was much too large to fit within the human body; rather, it was external and connected to the heart with vacuum tube catheter electrodes. The first human received a pacemaker implant in 1958. Ironically, Hopps himself received his own life-saving pacemaker in 1984 (Grant, 2017). And since its invention, it has saved countless lives.

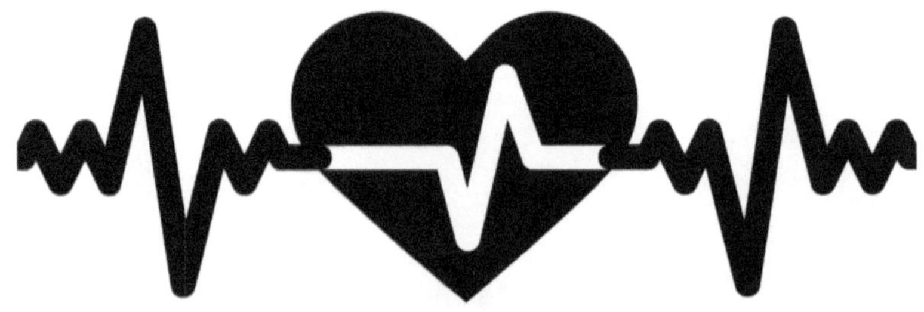

2. Insulin Process

Another life-saving discovery was the processing of insulin. Insulin is naturally occurring in the human body and is vital to the digestion and processing of sugars. However, those who live with type 1 diabetes produce little to no insulin, which causes sugars to build up in the body ("What is diabetes? n.d.). Today, the only way to counteract type 1 diabetes is through insulin injections. None of which would be possible if it weren't for the discovery of Dr. Frederick Banting and his team (Charles Best, J.J.R. Macleod, and Dr. Janes Collip) ("Our History," n.d.). Discovered in 1921, the first human patient received insulin treatments in January 1922 at Toronto General Hospital ("Our History," n.d.). By May 30, 1922, the University of Toronto and Eli Lilly & Co. of Indianapolis had reached a deal that allowed the mass production of insulin, making it available to the masses ("Our History," n.d.). And the rest is history, with October 31 of this year marking the 98th anniversary of Banting's original idea for insulin ("Our History," n.d.).

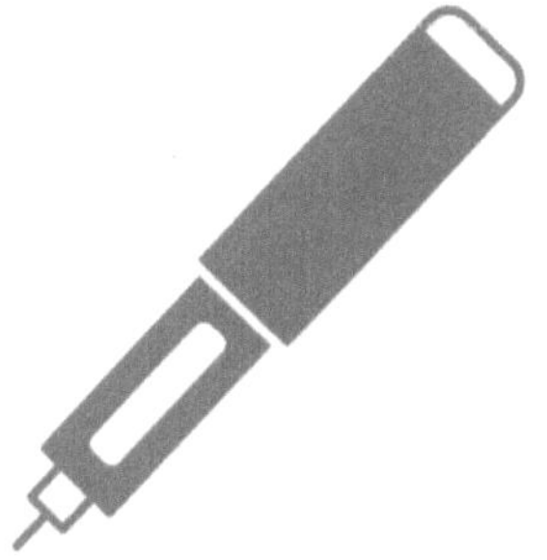

3. Film Colorization

Before 1983, there was no way to add colour to the classic black-and-white films that many had come to love. That is, until Canadian Wilson Markle was brought in to optimize footage from the Apollo missions. And it was his company, Image Transform, that converted the missions' footage to broadcast television (Martin, 2013). From there, Markle moved over to the movie industry to help cater to the demand for colour movies. And in 1983, Markle and his company, Colorization Inc., applied for the first of many patents for their frame-by-frame technique of colouring films. Of course, this was not without controversy, as many directors adamantly expressed their wish to keep their films in their original state of black and white (Martin, 2013). At the time of his retirement, Wilson Markle and his company had converted nearly 70 films, including It's a Wonderful Life, from black and white to colour, with him stating, "We were busy until there was very little black and white left to colour" (Martin, 2013).

4. BlackBerry and Smartphones

Who here has a smartphone? Well, considering nearly 50% of our readers tune in on their smartphones, I would say a large number of you do. I bet you didn't know that the first smartphone came from Canada. It's true. Mike Lazaridis and Doug Fregin founded their company, Research In Motion, in 1984 in Waterloo, Ontario (Leung, 2012). By 1999, this up-and-coming tech company had released its first device, a pager with email capabilities (Leung, 2012). After three short years and a name change, BlackBerry Limited released its very first smartphone in 2002 (Leung, 2012). Although the smartphone market has since shifted to Apple, Samsung, LG, and others, BlackBerry will always be the company that started it all.

5. Canadarm

Canada in Space is widely and primarily known for this piece of innovation and its follow-up, the Canadarm2, and later Dextre, which is

Canadian as well. Sorry, Chris Hadfield, you still hold a special place in all Canadians' hearts as the first Canadian astronaut to accomplish a spacewalk. But back to the Canadarm. During construction, the Canadarm proved to be a representation of Canadian ingenuity. Weighing in at 410 kg, it was actually unable to support itself while on Earth due to gravity, so all tests, verifications, and training had to be done via computer software (Doetsch & Lindberg, 2015). Deployed in 1981, the Canadarm flew 90 times on missions repairing and deploying several satellites (including Hubble), docking the space shuttle to the Russian Mir space station, removing ice from the shuttle, and helping construct the International Space Station (Doetsch & Lindberg, 2015). Capable of lifting over 266,000 kg in Space, the Canadarm was joined by the Canadarm2 in 2002, allowing both to work together in tandem (often nicknamed the "Canadian handshake") (Doetsch & Lindberg, 2015). In 2008, Dextre, a robot handyman capable of conducting tasks deemed unsafe for astronauts, joined the ranks and often worked in conjunction with the Canadarm2 (Doetsch & Lindberg, 2015). As of July 2011, Canadarm was retired after 30 years of operation (Doetsch & Lindberg, 2015).

And that's just the tip of the iceberg, with many other innovations coming from Canada, including IMAX, the walkie-talkie, the electric oven, the lightbulb, the A.M. radio, and sonar, to name a few more. With loads of new technology coming out each day, it's easy to forget where we've come from and how far we've come. And how those who have contributed to the scheme of things, including the many immigrants who have helped make their new homeland a better place; in fact, many of the inventions mentioned today were created thanks to many new Canadians. Something to keep in mind with the things going on in the world today. Well, that's all from us today at Schrodinger's Cat. Remember, Keep Looking for the Science of Things.

Driving on Sunshine: An Analysis of Solar Roads

(Originally Published September 18, 2018)

Since posting about the possibility of powering Canadian metropolitan areas with solar energy, we've received lots of positive feedback as well as one resounding question, "What's the deal with these solar roads?" We see it on the news, with tech giants introducing solar pathways and roadways onto their campuses. But are they feasible to power an entire city? And would they be any greener than the cities' current power source? In today's post, we look into the possibility of introducing solar roadways to a Canadian metropolitan area, the Greater Vancouver Area.

Canadian Weather

The first thing a city should consider is its local weather. In Canada, we get all four seasons (some are longer than others), and the Winter months can be detrimental to solar pathways and roadways. If they are covered in snow and ice, they are not converting solar energy at their optimum potential (if at all). Like in our previous post, some cities such as Calgary, Winnipeg, and Montreal, that are known to get large amounts of snow may find that installing solar panels would be a costly endeavour that reaps too few benefits to offset the cost. How do you keep them clear? Can snow shovels, gravel, and salt damage them? Are they more prone to ice buildup, thus causing a greater slipping hazard? These are just some of the questions that a city would have to ask.

Is It Possible in Canada?

One area that may be worth considering is BC's Lower Mainland which, for the most part, stays above freezing year-round. On average, the Lower Mainland gets around 9 days of snow a year, which averages around 44.6 centimetres (Osborn, n.d.). This is far more manageable than Montreal's average 59 days of snow, with an average snowfall of 209.5 centimetres (Osborn, n.d.). And as much as Vancouver is known for its rain, it gets about 1938 hours of sunshine each year, with an average of 41% sun ("Average Sunshine a Year in British Columbia," n.d.).

So from this information, we can determine that if solar panels were placed in Vancouver, they would receive:

$1370 \ W/m^2 \ x \ (0.41) = 561.7 \ W/m^2$

$561.7 \ W/m^2 \ x \ 1938 \ hours \ x \ (3600 \ seconds/hour)$
$= 3.92 \ x \ 10^9 \ J/m^2 \ in \ a \ given \ year.$

Do We Have Space?

The next thing that a city would need to consider is the amount of space they have to dedicate to this project. According to Patrick Johnstone's How Much Road is Enough?, there are approximately 19570 hectares $(1.96 \ x \ 10^8 \ m^2)$ in the Lower Mainland that have been converted into roadways (Johnstone, 2014).

Anmore = 3000 x 0.01 = 30 hectares
Bowen Is = 9000 x 0.03 = 270 hectares
Pitt Meadows = 5000 x 0.03 = 150 hectares
Belcarra = 1000 x 0.04 = 40 hectares
N. Van District = 16000 x 0.04 = 640 hectares
Maple Ridge = 27000 x 0.03 = 810 hectares
Langley = 31000 x 0.06 = 1860 hectares
Surrey = 32000 x 0.12 = 3840 hectares
Delta = 18000 x 0.09 = 1620 hectares
Coquitlam = 12000 x 0.08 = 960 hectares
Richmond = 13000 x 0.12 = 1560 hectares
W. Van = 9000 x 0.08 = 720 hectares
Port Moody = 3000 x 0.09 = 270 hectares
Lions Bay = 1000 x 0.11 = 110 hectares
PoCo = 4000 x 0.14 = 560 hectares

Burnaby = 9000 x 0.16 = 1440 hectares
Langley City = 2000 x 0.17 = 340 hectares
N. Van City = 2000 x 0.22 = 440 hectares
White Rock = 1000 x 0.25 = 250 hectares
New West = 2000 x 0.27 = 540 hectares
Vancouver = 12000 x 0.26 = 3120 hectares

Total Roadway Area = 19570 hectares = 1.96×10^8 m^2

From here, we can quickly determine the approximate amount of the Sun's energy that hits the roadways in the Lower Mainland (this does not take into account the amount of shade produced by skyscrapers) in a given year:

3.92×10^9 *J/m2 x (1.96×10^8 m^2)*
= 7.68×10^{17} J per year in the Lower Mainland

Meeting the Demand

Now, can it power the entire area? According to the Homeowners' Protection Office, the average amount of energy consumed in a given year in the Lower Mainland is around 213 kWh/m^2 (Finch et al., n.d.). So with the Lower Mainland having a total area of 212,000 hectares (2.12×10^9 m^2), you would need:

213 kWh/m^2 x 2.12×10^9 m^2 = 4.52×10^{11} kWh
= 1.63×10^{18} J needed per year

It's not looking good for solar energy. Next, we need to factor in the efficiency of solar panels. According to SolarRoadways, Sunpower Labs, a manufacturer of solar panel roadways, has solar roadways that are 18.5% efficient ("The Numbers," n.d.). So the total amount of energy they produce would be:

7.68×10^{17} J x (0.185) = 1.42×10^{17} J per year

This is clearly not enough; in fact, it would only produce enough energy to cover:

$$1.42 \times 10^{17} \, J \, / \, 1.63 \times 10^{18} \, J = 0.087$$
$$= \textit{around 9\% of the energy demand of the Lower Mainland.}$$

Is It Greener?

But what if we used that 9% to offset the current energy emissions? Vancouver is currently known as one of the greenest cities in North America when it comes to the amount of CO_2e (CO2 equivalent) emissions, and it has to do with its current energy provider: hydroelectricity (Economist Intelligence Unit, 2011). Hydroelectricity has had a reputation for being a fairly green energy source compared to other energy sources, such as coal and natural gas. However, its "cleanliness" depends on its efficiency, the age of its machinery, and its construction, among other factors. Well-maintained hydroelectric dams are capable of only producing around 18 tonnes of CO_2e/GWh, whereas the photoelectric cells of solar panels produce around 39 tonnes of CO_2e/GWh (Intrinsik Corp., 2016). To compare, coal and natural gas produce around 1,041 CO_2e/GWh and 622 CO_2e/GWh, respectively (Intrinsik Corp., 2016).

So is solar energy realistic for powering Vancouver? Not really. You could never produce enough solar energy to power the area, and you wouldn't really be reducing the CO_2e emissions any more than they already are. However, that doesn't mean that other cities shouldn't consider them. Cities in better climates, with more days and higher percentages of sunshine, as well as those that use "dirtier" energy sources, such as natural gas and coal, could see some benefits from solar energy or to even use it as a supplemental energy source. Also, a city such as Vancouver could consider solar roadways to power smaller needs, such as electric cars. That's all from us for now. Remember: Keep Looking for the Science in Things.

Footnote:

In December 2016, the first trial solar road was unveiled in Tourouvre, France. It was a 1-kilometre stretch of photovoltaic panels that were meant to generate electricity in order to power the streetlights of the town. However, the $5.6 million USD project was very short-lived. In May 2018, part of the road had to be deconstructed due to wear and tear. Another project near Caen, France, reportedly broke due to being struck by lightning. There were also apparent issues with leaves falling and obstructing the panels, as well as panels cracking as tractors drove on them, as they were not designed to withhold that weight. Is this the last time we see solar roads? Probably not. However, there are a lot of bugs that need to be worked through the system before we see them outside of trials.

6 Science Apps That You Need On Your Phone

(Originally Published October 9, 2018)

*In this post, we discuss some science apps that are our favourites, and bonus, they are FREE. We have not received any compensation (in the form of money or any other considerations) from anyone regarding
featuring and our opinions of these apps. These are our honest opinions, based on actual apps that we do use.*

Nowadays, there's an app for everything. Games, social media, news, health, and even dating. There are even science apps for everyone, from the everyday science enthusiast to those that want to learn something to university students that need help studying difficult concepts. So, today's post, we're going to discuss some of our favourite science apps that we currently have on our phones and that you may want to download if you haven't already.

Night Sky

If you don't have the Night Sky app on your phone, what are you doing? If you spend any time staring up at the night sky, you will want to add this app to your phone. Not only does the app pinpoint the notable constellations and planets overhead, it also shows meteors and satellites travelling overhead. Also included in the free version are the AR Sky Blending and World Traveller features, among many others. AR Sky Blending uses the camera on your phone and allows you to pull constellations out of the sky and explore them in more detail. World Traveller lets you put any location on Earth and allows you to see the night sky in that location. Curious about the Moon? The app also has a detailed map of the Moon, complete with area and crater names. So you can find out which craters make up the face of the "Man in the Moon." There are also premium features for those that wish to unlock additional features; however, you can still get great enjoyment out of the app in the free version.

NASA Visualization Explorer

If you love space, this is one app that you should check out. Available for both iOS and Android, NASA Visualization Explorer brings you the latest news from NASA and their research in the form of articles, videos, and stunning photos and imagery. Not only do they cover stories of their research in our solar system and beyond, but you can also see some of the research focused on our planet, including climate science, weather systems, and atmospheric studies. No need to worry about articles full of confusing jargon; this app presents its articles in short and concise descriptions that you can download for offline viewing and read on your coffee break at work with plenty of time to spare.

PalmOil Scan

If you want to minimize your impact on the environment and the planet, you will want to add this app to your phone. Palm oil production has become a cause for concern in recent years due to the traditionally unsustainable practises used to produce it, which result in the destruction of vast tracts of rainforest, the endangerment of many animals, and the loss of areas known as carbon sinks (which absorb atmospheric carbon dioxide and convert it into oxygen). And the demand for palm oil is not going away. If you take a look around your house, the number of products containing palm oil is startling. So how can you tell if the products you are buying contain palm oil that is sustainably produced? Look no further than this app. Created in conjunction with the Cheyenne Mountain Zoo, this app allows you to scan the barcode of the product or perform a quick product search of the item that you are questioning, and it quickly informs you if your product contains sustainable palm oil or not. This app is available on both iOS and Android and allows consumers to make informed decisions and exercise their purchasing power to make a positive environmental impact.

Think Dirty

It seems every day something new in your house is trying to kill you, and unless you check all the ingredients in all the products around your house, you'll never know what you are exposing yourself to. It sounds like a long and tedious process, doesn't it? Not with the Think Dirty app. Available on Android and iOS, you scan the barcode of any product in your house that you are wondering about, and it will give the chemical "cleanliness" score, highlighting any concerning ingredients, their usages, and the health impacts they cause. So if you are concerned about the chemicals that you are exposing yourself to, you may want to add this app to your phone.

iNaturalist & Seek by iNaturalist

If you love getting out into nature and enjoying its beauty, these apps may be for you. iNaturalist helps you identify organisms that you may encounter on your adventures. Simply take a picture using the app, and it will generate suggestions based on your photo and location, so you can accurately identify the organism and keep tabs on all your sightings.

Want to take your discoveries up a notch? Seek lets you know what organisms people are finding in your area, so you can search for them. Also, with Seek, the more images you collect, the more badges you earn. Make your next hike or nature walk a fun scavenger hunt that you can enjoy with every member of your family. iNaturalist's identification app is available on iOS and Android, and Seek is available on iOS (they are currently working on developing an Android version; in the meantime, Android users can use the 'Mission" feature of the iNaturalist app).

In our day-to-day lives, it is easy to get wrapped up in our phones and the apps that we have on them. And if you can't cut the cellular addiction (I've been told that my phone is my appendage sometimes), you might as well add some science apps and increase your know-how. Some science apps can even help you get out in nature and learn something new about the world around you. In the meantime, Keep Looking for the Science in Things.

Works Cited

The Chemistry of Cannabis

Hillig, Karl W, and Paul G Mahlberg. "A chemotaxonomic analysis of cannabinoid variation in Cannabis (Cannabaceae)." *American Journal of Botany* vol. 91,6 (2004): 966-75.

"Inside: Marijuana Facts." *National Geographic TV*, https://natgeotv.com/ca/inside-marijuana/facts. Accessed June 2018.

NIDA. "Marijuana." *National Institute on Drug Abuse*, https://www.drugabuse.gov/publications/drugfacts/marijuana. Accessed June 2018.

Rafael-Hawkins, Olivia. "Indica vs. Sativa: What's The Difference?" *Leaf Science*, 16 October 2017, https://www.leafscience.com/2017/10/16/indica-vs-sativa-whats-difference/. Accessed June 2018.

Research and Statistics Division. "Cannabis crime statistics in Canada, 2020." *Just Facts,* June 2022, https://www.justice.gc.ca/eng/rp-pr/jr/jf-pf/2022/pdf/RSD_JF2022_Cannabis-related-crime-statistics2020-en.pdf#:~:text=Following%20its%20legalization%2C%20cannabis-related%20drug%20offences%20made%20up,a%20rate%20of%20176%20offences%20per%20100%2C000%20population.. Accessed June 2023.

The Canadian Press. "Cannabis has added $43.5B to Canada's economy since legalization: report." *Global News,* 1 February 2022, https://globalnews.ca/news/8585983/cannabis-canada-economy-legalization-report/. Accessed June 2023.

Vandergriendt, Carly. "Sensation of a Marijuana High: Smoking, Edibles, and Vaping." *Healthline,* 7 Mar 2019, https://www.healthline.com/health/what-does-it-feel-like-to-be-high. Accessed 30 Dec 2022 [To replace a citation from the original blog post that is no longer accessible].

The Science Behind Fireworks

Cohen, Jennie. "Fireworks' Vibrant History." *History*,
http://www.history.com/news/fireworks-vibrant-history. Accessed June 2018.

Dematto, Amanda. "The Secret Science Inside Brilliant Fireworks." *Popular Mechanics*, 02 Jul 2010,
https://www.popularmechanics.com/technology/a5911/how-fireworks-are-made/.
Accessed June 2018.

"Fireworks." *Wikipedia,* https://en.wikipedia.org/wiki/Fireworks. Accessed June 2018.

The Colours of the Wind

Bassham, James Alan, and Hans Lambers. "Photosynthesis." *Encyclopedia Britannica*,
https://www.britannica.com/science/photosynthesis. Accessed September 2018.

Palm Jr., Carl E. "Why Leaves Change Color." *ESF: SUNY College of Environmental Science and Forestry: Around Your World*,
https://www.esf.edu/pubprog/brochure/leaves/leaves.htm. Accessed September 2018.

Bet You Didn't Know That Was A GMO

"Dichlorodiphenyltrichloroethane (DDT)." *Centers for Disease Control and Prevention,* 16 August 2021,
https://www.cdc.gov/biomonitoring/DDT_FactSheet.html. Accessed June 2023.

Freedman, David H. "The Truth about Genetically Modified Food." *Scientific American*, 1 Sept 2013,
https://www.scientificamerican.com/article/the-truth-about-genetically-modified-food/. Accessed April 2018.

Lallanilla, Marc. "GMOs: Facts About Genetically Modified Food." *LiveScience,* 11 Jan 2016, https://www.livescience.com/40895-gmo-facts.html. Accessed April 2018.

"Monsanto." *Encyclopedia Britannica,* 3 May 2023, https://www.britannica.com/topic/Monsanto. Accessed June 2023.

Neslen, Arthur. "Monsanto sold banned chemicals for years despite known health risks, archives reveal." *The Guardian,* 9 August 2017, https://www.theguardian.com/environment/2017/aug/09/monsanto-continued-selling-pcbs-for-years-despite-knowing-health-risks-archives-reveal. Accessed June 2023.

Olby, Robert. "Gregor Mendel." *Encyclopedia Britannica,* https://www.britannica.com/biography/Gregor-Mendel. Accessed April 2018.

Olson, Eric R. "What is a Genetically Modified Food?" *Scientific American,* 5 Jan 2015,

https://www.scientificamerican.com/video/what-is-a-genetically-modified-food2013-07-24/. Accessed April 2018.

Zhang, Chen, Wohlhueter, Robert, and Han Zhang. "Genetically modified foods: A critical review of their promise and problems." *Food Science and Human Wellness* vol. 5,3 (2016): 116-123.

10 Endangered Animals Brought Back From The Brink

"American Alligator: *Alligator Misissippiensis.*" *The Nature Conservancy,* http://www.nature.org/newsfeatures/specialfeatures/animals/reptiles/american-alligator.xml. Accessed May 2018.

"American Alligator: *Alligator mississippiensis.*" *Florida Fish and Wildlife Conservation Commission,* https://myfwc.com/wildlifehabitats/profiles/reptiles/alligator/. Accessed 2 Jan 2023.

Beans, Laura. "10 Success Stories Thanks to the Endangered Species Act." *EcoWatch,* 07 Dec 2013, https://www.ecowatch.com/10-success-stories-thanks-to-the-endangered-species-act-1881837279.html. Accessed May 2018.

"Greater One-Horned Rhino: Facts." *WWF,* https://www.worldwildlife.org/species/greater-one-horned-rhino. Accessed May 2018.

"Green Turtle: Facts." *WWF*, https://www.worldwildlife.org/species/green-turtle. Accessed May 2018.

Hatfield, B.B., Yee, J.L., Kenner, M.C., and Tomoleoni, J.A. "California Sea Otter (Enhydra lutris nereis) Census Results, Spring 2019." *U.S. Geological Survey* Data Series 1118 (2019).

"Humpback Whale: Facts." *WWF*, https://wwf.panda.org/discover/our_focus/wildlife_practice/profiles/mammals/whales_dolphins/humpback_whale/. Accessed May 2018.

Long, Claudia. "Humpback whales no longer listed as endangered after major recovery." *ABC News*, 25 Feb 2022. https://www.abc.net.au/news/2022-02-26/humpback-whales-no-longer-listed-as-endangered/100862644?utm_campaign=abc_news_web&utm_content=link&utm_medium=content_shared&utm_source=abc_news_web. Accessed 2 Jan 2023.

Newsroom. "Indian Rhino numbers top 4,000." *Rhino Review*, 13 May 2022, https://rhinoreview.org/indian-rhino-numbers-top-4000/. Accessed 30 Dec 2022.

Kotala, Zenaida. "Green Sea Turtles Break Nesting Record on Florida Beaches." *UCF Today*, 2 Sept 2015, https://www.ucf.edu/news/green-sea-turtles-break-nesting-record-on-florida-beaches/. Accessed May 2018.

Middle Island Warrnambool, http://www.warrnamboolpenguins.com.au/index.php. Accessed May 2018.

"Population." *Save The Manatee*, https://www.savethemanatee.org/manatees/manatee-population/. Accessed 2 Jan 2023.

Pritchard AM, Sanchez CL, Bunbury N, Burt AJ et al. "Green turtle population recovery at Aldabra Atoll continues after 50 yr of protection." *Endangered Species Research*, vol 47 (2022): 205 -215.

"Southern Sea Otter." *U.S. Fish & Wildlife Services: Ventura Fish and Wildlife Office*, https://www.fws.gov/ventura/endangered/species/info/sso.html. Accessed May 2018.

Smith, Larissa et al. "New Jersey Bald Eagle Project, 2015." *New Jersey Department of Environmental Protection: Division of Fish and Wildlife,* http://www.conservewildlifenj.org/downloads/cwnj_676.pdf. Accessed May 2018.

"Statewide Nuisance Alligator Program." *Florida Fish and Wildlife Conservation Commission,* https://myfwc.com/wildlifehabitats/wildlife/alligator/snap/. Accessed 2 Jan 2023.

"Trumpeter Swan," *Animalia,* https://animalia.bio/trumpeter-swan#:~:text=According%20to%20Wikipedia%20sources%2C%20the%20Trumpeter%20swan%20population,least%20concern%20%28LC%29%20on%20the%20IUCN%20Red%20List.. Accessed 30 Dec 2022.

"Trumpeter swans continue to be a conservation success story." *Department of Natural Resources Minnesota,* 11 Feb 2016, http://news.dnr.state.mn.us/2016/02/11/trumpeter-swans-continue-to-be-a-conservation-success-story/. Accessed May 2018.

U.S. Fish and Wildlife Service. "Eagle Permits; Updated Bald Eagle Population Estimates and Take Limits." *Federal Register,* 1 Feb 2022, https://www.federalregister.gov/documents/2022/02/01/2022-02040/eagle-permits-updated-bald-eagle-population-estimates-and-take-limits. Accessed 2 Jan 2023.

Watson, Traci. "Rare Butterflies Flying High at Los Angeles Airport." *National Geographic,* 21 April 2016, http://news.nationalgeographic.com/2016/04/160421-butterflies-endangered-species-animals/. Accessed May 2018.

"West Indian Manatee: Species Profile." *National Park Service,* 3 June 2021, https://www.nps.gov/ever/learn/nature/manateepage.htm. Accessed 30 Dec 2022 [To replace a citation from the original blog post that is no longer accessible].

Un-Bee-lievable

Bennet, Lauren. "7 Species of Bees Added to Endangered List." *Climate Institute,* 6 Mar 2016, http://climate.org/7-species-of-bees-added-to-endangered-list. Accessed June 2018.

Bessin, Ric. "Varroa Mites Infesting Honey Bee Colonies." *Entomology at the University of Kentucky*, April 2016, http://entomology.ca.uky.edu/ef608. Accessed June 2018.

Boisvert, Nick. "How DNA could be key to stopping the dangerous decline of Canada's bee colonies." *CBC News*, https://www.cbc.ca/news/canada/toronto/honey-bee-genomics-project-1.5288952. Accessed June 2023.

Carrington, Damian. "EU agrees total ban on bee-harming pesticides." *The Guardian*, 27 April 2018, https://www.theguardian.com/environment/2018/apr/27/eu-agrees-total-ban-on-bee-harming-pesticides. Accessed June 2023.

Daftardar, Ishan. "Why Bee Extinction Would Mean the End of Humanity." *Science ABC*, 3 Jul 2015, https://www.scienceabc.com/nature/bee-extinction-means-end-humanity.html. Accessed June 2018.

Deeley, Anita. " A beginner beekeeper's guide to the parts of a beehive." *Beverly Bees*, https://www.beverlybees.com/parts-beehive-beginner-beekeeper/. Accessed June 2018.

Hopwood, Jennifer. "Effects of Neonicotinoid Insecticides on Bees." *The Xerces Society for Invertebrate Conservation*, https://jhawkins54.typepad.com/files/effects_of_neonicotinoid_insecticides_on_bees-hopwood.pdf. Accessed June 2018.

Hopwood, Jennifer, Code, Aimee, Vaughan, Mace, Biddinger, David. Shepherd, Matthew, Hoffman Black, Scott, Lee-Mäder, Eric, and Celeste Mazzacano. "How Neonicotinoids Can Kill Bees: The Science Behind the Role These Insectides Play in Harming Bees." *The Xerces Society for Invertebrate Conservation*, 2016.

Liao, Kristine. "Cities are letting plants go wild for 'No Mow May.'" *Popular Science*, 10 May 2022, https://www.popsci.com/environment/no-mow-may-environmental-movement/. Accessed June 2023.

Moate, Maddie. "What would happen if bees went extinct?" *Earth Unplugged*, 4 May 2014, http://www.bbc.com/future/story/20140502-what-if-bees-went-extinct. Accessed June 2018.

Nace, Trevor. "Morgan Freeman Converted His 124-Acre Ranch Into A Giant Honeybee Sanctuary To Save The Bees." *Forbes,* 20 May 2019, https://www.forbes.com/sites/trevornace/2019/03/20/morgan-freeman-converted-his -124-acre-ranch-into-a-giant-honeybee-sanctuary-to-save-the-bees/?sh=ae524badfa55. Accessed June 2023.

Spivak, Marla. "Why bees are disappearing." *TED*, 17 Sep 2013, https://www.youtube.com/watch?v=dY7iATJVCso&feature=youtu.be. Accessed June 2018.

Zattara, Eduardo E., and Marcelo A. Aizen. "Worldwide occurrence records suggest a global decline in bee species richness." *OneEarth*, vol. 4, 1(2021): 114-123.

They Are Taking Over: Aquatic Invasive Species

"FLOWERING RUSH Distribution Map, As of September 2020." *Government of Alberta,*

https://open.alberta.ca/dataset/080f0ec8-e41a-422d-862f-e549e4087456/resource/cec 6f444-df0d-4470-9256-aff6f0357d50/download/aep-flowering-rush-distribution-map-2021.pdf. Accessed June 2023.

"Invasive Mussel Detection Project." *Working Dogs for Conservation,* https://wd4c.org/our-work-biosecurity-invasives/invasive-mussel-detection-project. Accessed June 2023.

"Quagga Mussel." *Fisheries and Oceans Canada: Aquatic Invasive Species,* 22 June 2020, https://www.dfo-mpo.gc.ca/species-especes/profiles-profils/quaggamussel-moulequagg a-eng.html. Accessed June 2023.

"Zebra and Quagga Mussels: *Dreissena polymorpha & Dreissena bugensis*" *Invading Species*, http://www.invadingspecies.com/zebra-quagga-mussels/. Accessed July 2018.

"Zebra Mussel." *Fisheries and Oceans Canada: Aquatic Invasive Species*, 26 March 2021,
https://www.dfo-mpo.gc.ca/species-especes/profiles-profils/zebramussel-moulezebree-e
ng.html#distribution. Accessed June 2023.

Diving Into the World of Sharks

Capuzzo, Michael. "The true story of Jaws." *BBC Culture*, 13 July 2016,
https://www.bbc.com/culture/article/20160713-the-true-story-of-jaws. Accessed July
2018.

"Incidents by country." *Shark Attack Data*, http://www.sharkattackdata.com/place.
Accessed July 2018.
Martin, R. Aidan. "Shark Evolution: Fathoming Geologic Time." *Biology of Sharks
and Rays*, http://www.elasmo-research.org/education/evolution/geologic_time.htm.
Accessed July 2018.

"Shark: Facts." *WWF*, https://www.worldwildlife.org/species/shark. Accessed July
2018.

"Sharks and Rays," *Defenders of Wildlife*, https://defenders.org/sharks/basic-facts.
Accessed July 2018.

The Ocean Portal Team. "Sharks: Euselachii." *Smithsonian*,
https://ocean.si.edu/ocean-life/sharks-rays/sharks. Accessed July 2018.

University of Florida. "Shark Biology." *Florida Museum*,
https://www.floridamuseum.ufl.edu/fish/discover/sharks/biology/. Accessed July
2018.

"Whale Shark." *National Geographic*,
https://www.nationalgeographic.com/animals/fish/w/whale-shark/. Accessed July
2018.

WHO. "Global Health Observatory (GHO) data." *World Health Organization*,
https://www.who.int/gho/road_safety/mortality/en/. Accessed July 2018

Left-Handed, Right-Handed

Brandler, William M., Morris, Andrew P., Evans, David M., Scerri, Thomas S., Kemp, John P., Timpson, Nicholas J., St Pourcain, Beate, Smith, George Davey, Ring, Susan M., Stein, John, Monaco, Anthony P., Talcott, Joel B., Fisher, Simon E., Webber, Caleb, and Silvia Paracchini. "Common Variants in Left/Right Asymmetry Genes and Pathways Are Associated with Relative Hand Skill." *PLOS GENETICS* vol. 9, 9 (2013): e1003751.

Cole, J. "Paw preference in cats related to hand preference in animals and men." *Journal of Comparative and Physiological Psychology* vol. 48, 2 (1955): 137 - 140.

Cook, Sharell. "12 Indian Etiquette Don'ts." *TripSavvy*, https://www.tripsavvy.com/indian-etiquette-donts-1539435. Accessed August 2018.

Dragovic, M., and G. Hammond. "Handedness in schizophrenia: a quantitative review of evidence." *Acta Psychiatrica Scandinavica* vol. 111, 6 (2005): 410 - 419.

Finch, G. "Chimpanzee handedness." *Science* vol. 94 (1941): 117 - 118.

Geilig, Natasha. "Why Are Some People Left-Handed?" *Ask Smithsonian,* 12 Sept 2013, https://www.smithsonianmag.com/science-nature/why-are-some-people-left-handed-6556937/. Accessed August 2018.

Klar, Amar J. S. "Human Handedness and Scalp Hair-Whorl Direction Develop From a Common Genetic Mechanism." *GENETICS* vol. 165, 1 (2003): 269 - 276.

Knecht, S., Dräger, B., Deppe, M., Bobe, L., Lohmann, H., Flöel, A., Ringelstein, E.-B., and H. Henningsen. "Handedness and hemispheric language dominance in healthy humans." *Brain* vol. 123, 12 (2000): 2512 - 2518.

Kushner, Howard I. "Why are there (almost) no left-handers in China?" *Endeavour* vol. 37, 2 (2013): 71 - 81.

"Left and Right Hemispheres." *The Brain Made Simple,* http://brainmadesimple.com/left-and-right-hemispheres.html. Accessed Aug 2018.

Mastin, Luke. "Other Handedness Issues - Handedness and Animals." *Right Left Right Wrong?* http://www.rightleftrightwrong.com/issues_animals.html. Accessed August 2018.

Sommer, Iris, Aleman, André, Ramsey, Nick, Bouma, Anke, and René Kahn. "Handedness, language lateralisation and anatomical asymmetry in schizophrenia." *British Journal of Psychiatry* vol. 178, 4 (2001): 344 - 351.

Tan, Üner. "Paw Preferences in Dogs." *International Journal of Neuroscience* vol. 32, 3 - 4 (1987): 825 - 829.

Yoshioka, Joseph G. "Handedness in Rats." *The Pedagogical Seminary and Journal of Genetic Psychology* vol. 38, 1 (1930): 1 - 4, 471 - 474.

Turkey Coma: Fact or Scapegoat?

Ballantyne, Coco. "Does Turkey Make You Sleepy?" *Scientific American*, 21 Nov 2007. https://www.scientificamerican.com/article/fact-or-fiction-does-turkey-make-you-sleepy/. Accessed Sept 2018.

Zamosky, Lisa. "The Truth About Tryptophan." *WebMD*, 18 Nov 2009, https://www.webmd.com/food-recipes/features/the-truth-about-tryptophan. Accessed Sept 2018.

The Great Avian Migration

Connor, Steve. "Revealed: secret of how birds navigate during migration." *Independent*, 1 May 2008, https://www.independent.co.uk/news/science/revealed-secret-of-how-birds-navigate-during-migration-818766.html. Accessed Oct 2018.

Mouly, Ann-Marie, and Regina Sullivan. "Memory and Plasticity in the Olfactory System: From Infancy to Adulthood." *The Neurobiology of Olfaction*, edited by Menini, Boca Raton, FL, CRC Press/Taylor and Francis, 2010, www.ncbi.nlm.nih.gov/books/NBK55967.

It Came From The Deep: The Headless Chicken Monster

Miller, John E., and David Leo Pawson. "Echinoderm." *Encyclopedia Britannica*, https://www.britannica.com/animal/echinoderm. Accessed Oct 2018.

"Sea cucumber." *Encyclopedia Britannica*, https://www.britannica.com/animal/sea-cucumber. Accessed Oct 2018.

4 Ways to Make the Planet Greener, While Saving You Some Green

"Global Warming and Your Car." *Car Talk*, https://www.cartalk.com/content/global-warming-and-your-car-0. Accessed April 2018.

"How Energy-Efficient Light Bulbs Compare with Traditional Incandescents." *Energy Saver*, https://www.energy.gov/energysaver/save-electricity-and-fuel/lighting-choices-save-you-money/how-energy-efficient-light. Accessed April 2018.

"Learn about CFLs." *Energy Star*, https://www.energystar.gov/products/lighting_fans/light_bulbs/learn_about_cfls. Accessed April 2018.

Taylor, Glenda. "How To: Dispose of Light Bulbs." *Bob Vila*, https://www.bobvila.com/articles/how-to-dispose-of-light-bulbs/. Accessed April 2018.

Wakeland, Wayne, Cholette, Susan and Kumar Venkat. "Food transportation issues and reducing carbon footprint." http://www.cleanmetrics.com/pages/ch9_0923.pdf. Accessed April 2018.

"Water Questions & Answers: How much water does the average person use at home per day?" *USGS*, https://water.usgs.gov/edu/qa-home-percapita.html. Accessed April 2018.

The Lowdown on Wind Turbines

Appendix J: *V90 3.0MW Turbine Specifications*, Vestas, http://www.gov.pe.ca/photos/original/eef_epwf_turbin.pdf. Accessed June 2023.

"Average Annual Wind Speed at Canadian Cities." *Current Results*, https://www.currentresults.com/Weather/Canada/Cities/wind-annual-average.php. Accessed May 2018.

"Efficiency of Wind Turbines." *Linquip Technews,*
https://www.linquip.com/blog/efficiency-of-wind-turbines/. Accessed June 2023.

Power curve graph, *wind-turbine-models.com,*
https://en.wind-turbine-models.com/turbines/16-vestas-v90. Accessed June 2023.

"Wind Turbine Calculator," *WindCycle.energy,*
https://windcycle.energy/wind_turbine_calculator/#:~:text=Finding%20the%20efficie
ncy%20of%20the%20turbine%20You%20can,-%20kt%29%20%2A%20%281%20-%20k
w%29%20%2A%20C%E2%82%9A. Accessed June 2023.

Kilauea: The Bubbling Giant

Bagley, Mary. "Volcano Facts and Types of Volcanoes." *LiveScience*, 6 Feb 2018.
https://www.livescience.com/27295-volcanoes.html. Accessed May 2018.

Harrison, Christopher G. A. "The present-day number of tectonic plates." *Earth,
Planets and Space* vol 68, 1 (2016): 1.

"How did the Hawaiian Islands form?" *National Ocean Service,*
https://oceanservice.noaa.gov/facts/hawaii.html. Accessed 3 Jan 2023. [To replace a
citation from the original blog post that is no longer accessible].

"Kilauea: Eruptive History." *Smithsonian Institution | Global Volcanism Program,*
https://volcano.si.edu/volcano.cfm?vn=332010&vtab=Eruptions. Accessed June 2023.

"Kilauea Volcano." *Basic Planet,* http://www.basicplanet.com/kilauea-volcano/.
Accessed May 2018.

"Kilauea volcano." *Volcano Discovery,*
https://www.volcanodiscovery.com/kilauea.html. Accessed May 2018.

"Lava flows - can you outrun a lava flow?" *GEOetc: Geoscience Education & Outreach,*
https://geoetc.com/lava-flows-speed/. Accessed 3 Jan 2023. [To replace a citation from
the original blog post that is no longer accessible].

Lynch, David. "Volcanoes." *SanAndreasFault.org,*
http://sanandreasfault.org/Volcanoes.html#:~:text=The%20majority%20of%20the%20
world%27s%20volcanoes%20are%20associated,a%20great%20while%2C%20a%20new%
20volcano%20will%20form.. Accessed 3 Jan 2023. [To replace a citation from the
original blog post that is no longer accessible].

"Mauna Loa." *USGS: Volcano Hazards Program,* https://volcanoes.usgs.gov/volcanoes/mauna_loa/geo_hist_summary.html. Accessed May 2018.

Pfeiffer, Tom. "How Many Volcanic Eruptions Occur Every Year?" *Volcano Discovery,* https://www.volcanodiscovery.com/es/volcanology/faq/how-many-eruptions-per-year.html#:~:text=How%20many%20volcanic%20eruptions%20occur%20every%20year%3F%20On,is%20more%20in%20the%20range%20of%20about%2060-80.. Accessed 3 Jan 2023. [To replace a citation from the original blog post that is no longer accessible].

"Ring of Fire." *National Geographic,* https://education.nationalgeographic.org/resource/ring-fire. Accessed 3 Jan 2023. [To replace a citation from the original blog post that is no longer accessible].

Robertson, Eugene C. "The Interior of the Earth." *USGS,* 14 Jan 2011, https://pubs.usgs.gov/gip/interior/. Accessed May 2018.

"Types of Volcanoes." *Basic Planet,* 2 May 2013, https://www.basicplanet.com/types-of-volcanoes/. Accessed May 2018.

Volcano Hazards. "Pyroclastic flows move fast and destroy everything in their path." *USGS,* https://www.usgs.gov/programs/VHP/pyroclastic-flows-move-fast-and-destroy-everything-their-path. Accessed 3 Jan 2023. [To replace a citation from the original blog post that is no longer accessible].

"White Island volcanic eruption of 2019." *Encyclopedia Britannica,* 26 May 2023, https://www.britannica.com/event/White-Island-volcanic-eruption-of-2019. Accessed June 2023.

Hurricanes, Cyclones, Typhoons... Oh My!

Belles, Jonathan. "Hurricane vs. Tropical Storm: What's the Difference and What Does it Mean for Harvey?" *The Weather Channel,* 26 Aug 2017, https://weather.com/science/weather-explainers/news/tropical-storm-vs-hurricane-harvey. Accessed May 2018.

Brain, Marshall, Freundenrich, Craig, and Robert Lamb. "Lifecycle of a Hurricane." *HowStuffWorks.com,* 25 Aug 2000, https://science.howstuffworks.com/nature/natural-disasters/hurricane3.htm, Accessed May 2018.

"How do hurricanes form?" *NASA Space Place,* https://spaceplace.nasa.gov/hurricanes/en/. Accessed May 2018.

"How do hurricanes form?" *National Ocean Service,* https://oceanservice.noaa.gov/facts/how-hurricanes-form.html. Accessed 3 Jan 2023. [To replace a citation from the original blog post that is no longer accessible].

"Hurricane Safety Tips and Resources." *National Weather Service,* https://www.weather.gov/safety/hurricane. Accessed May 2018.

"Saffir-Simpson Hurricane Wind Scale." *National Hurricane Center,* https://www.nhc.noaa.gov/aboutsshws.php. Accessed May 2018.

"Storm Surge Overview," *National Hurricane Center and Central Pacific Hurricane Center,* https://www.nhc.noaa.gov/surge/. Accessed May 2018.

"Tropical Cyclone Climatology," *National Hurricane Center,* https://www.nhc.noaa.gov/climo/. Accessed May 2018.

"What is the difference between a hurricane and a typhoon?" *National Ocean Service,* https://oceanservice.noaa.gov/facts/cyclone.html. Accessed May 2018.

"When is hurricane season?" *Atlantic Oceanographic & Metrological Laboratory: Hurricane Research Division,* https://www.aoml.noaa.gov/hrd/tcfaq/G1.html. Accessed May 2018.

Schrodinger's Cat's Quick Guide to Climate Change

"As Canada bans bags and more, this is what's happening with single-use plastics around the world." *World Economic Forum*, 26 October 2020, https://www.weforum.org/agenda/2020/10/canada-bans-single-use-plastics/. Accessed July 2023.

Brain, Marshall, "If the polar ice caps melted, how much would the oceans rise?" *HowStuffWorks.com,* 21 Sept 2000,

https://science.howstuffworks.com/environmental/earth/geophysics/question473.htm
. Accessed May 2018.

Carlowicz, Michael, "Tracking 30 Years of Sea Level Rise." *Earth Observatory,*
https://earthobservatory.nasa.gov/images/150192/tracking-30-years-of-sea-level-rise.
Accessed July 2023.

Cline, William R. "Global Warming and Agriculture." *Finance & Development,* March
2008, http://www.imf.org/external/pubs/ft/fandd/2008/03/pdf/cline.pdf. Accessed
May 2018.

"Frequently asked questions about climate change." *Government of Canada,* 30 Nov
2015,
https://www.canada.ca/en/environment-climate-change/services/climate-change/frequ
ently-asked-questions.html. Accessed May 2018.

Geggel, Laura. "How Often Do Ice Ages Happen?" *LiveScience,* 25 March 2017,
https://www.livescience.com/58407-how-often-do-ice-ages-happen.html. Accessed
May 2018.

"Global Warming." *Earth Observatory,* 3 June 2010,
https://earthobservatory.nasa.gov/features/GlobalWarming/page2.php. Accessed May
2018.

"Greenhouse effect." *Australian Government: Department of the Environment and
Energy,*
http://www.environment.gov.au/climate-change/climate-science-data/climate-science/
greenhouse-effect. Accessed May 2018.

"Surface air temperature map." *The Copernicus Project,* July 2023,
https://climate.copernicus.eu/surface-air-temperature-maps. Accessed July 2023.

"Vital Signs: Global Temperature." *Earth Science Communication Team of NASA's Jet
Propulsion Laboratory,* 13 July 2023,
https://climate.nasa.gov/vital-signs/global-temperature/. Accessed July 2023.

Watkins, Thayer. "Snowball Earth: The Discovery of Evidence That the Earth Was
Once Frozen Over from Pole to Pole." *San Jose State University,*
http://www.sjsu.edu/faculty/watkins/snowball.htm. Accessed May 2018.

<u>6 Ocean Habitats You've Never Heard Of</u>

Andrews, Gianna. "Plastics in the Ocean Affecting Human Health." *Geology and Human Health*,
https://serc.carleton.edu/NAGTWorkshops/health/case_studies/plastics.html.
Accessed June 2018.

"Animals of the Ice: Antarctic Krill." *Ocean Today*,
https://oceantoday.noaa.gov/animalsoftheice_krill/. Accessed June 2018.

Chan, Melissa. "Oceans Will Have More Plastic Than Fish By 2050, Study Says." *Time*,
20 Jan 2016, http://time.com/4186250/ocean-plastic-fish. Accessed June 2018.

"Cold Seeps." *Project Oceanography*, 2002,
http://www.marine.usf.edu/pjocean/packets/sp02/sp02u2p4.pdf. Accessed June 2018.

"Deep Hydrothermal Vent." *Oceana*,
https://oceana.org/marine-life/marine-science-and-ecosystems/deep-hydrothermal-vent
. Accessed June 2018.

"Eel Suffers Toxic Shock From Brine Pool | Blue Planet II." *BBC Earth*, 3 Nov 2017,
https://www.youtube.com/watch?v=ZwuVpNYrKPY. Accessed June 2018.

"Fjord." *Oceana*, https://oceana.org/marine-life/marine-science-and-ecosystems/fjord.
Accessed June 2018.

"Garbage Patches: How Gyres Take Our Trash Out to Sea." *NOAA Ocean Podcast*,
https://oceanservice.noaa.gov/podcast/mar18/nop14-ocean-garbage-patches.html?trk=
public_post_comment-text. Accessed July 2023.

"How much of the ocean have we explored?" *National Ocean Service*,
https://oceanservice.noaa.gov/facts/exploration.html. Accessed June 2018.

"Krill Hotspots: About." *British Antarctic Survey: Natural Environment Research Council*, https://www.bas.ac.uk/project/krill-hotspots/. Accessed June 2018.

"Latest News." *Canadian Parks and Wilderness Society*, 23 January 2022,
https://glassspongereefs.com/latest-news/. Accessed July 2023.

"Seamounts." *Woods Hole Oceanographic Institution,*
http://www.whoi.edu/main/topic/seamounts. Accessed June 2018.

"Scientists Discover Three New Hydrothermal Vent Fields on Mid-Atlantic Ridge."
Schmidt Ocean Institute, 20 April 2023,
https://schmidtocean.org/scientists-discover-three-new-hydrothermal-vent-fields-on-m
id-atlantic-ridge/. Accessed July 2023.

"The Biology." *Canadian Parks and Wilderness Society,*
http://glassspongereefs.com/what-is-a-glass-sponge. Accessed June 2018.

"The Discovery." *Canadian Parks and Wilderness Society,*
https://glassspongereefs.com/the-discovery/. Accessed June 2018.

"The search for answers in the ice." *NIWA Taihoro Nukurangi*, 31 May 2022,
https://niwa.co.nz/news/the-search-for-answers-in-the-ice. Accessed July 2023.

"What is a seamount?" *National Ocean Service,*
https://oceanservice.noaa.gov/facts/seamounts.html. Accessed June 2018.

Shining A Light on Solar Energy

"Sunshine in Canadian Cities: Average Hours & Days a Year." *Current Results,*
https://www.currentresults.com/Weather/Canada/Cities/sunshine-annual-average.ph
p. Accessed June 2018.

Put Down The Plastic

Geyer, Roland, Jambeck, Jenna R., and Kara Lavander Law. "Production, use, and fate
of all plastics ever made." *Science Advances* vol. 3, 7 (2017): e1700782.

"How Is Plastic Recycled: Step by Step." *Greentumble*, 24 May 2018,
https://greentumble.com/how-is-plastic-recycled-step-by-step/. Accessed July 2018.

Laville, Sandra and Matthew Taylor. "A million bottles a minute: world's plastic binge
'as dangerous as climate change.'" *The Guardian,* 28 June 2017,
https://www.theguardian.com/environment/2017/jun/28/a-million-a-minute-worlds-
plastic-bottle-binge-as-dangerous-as-climate-change. Accessed July 2018.

Leblanc, Rick. "The Decomposition of Waste in Landfills: A Story of Time and Materials." *The Balance*, https://www.thebalancesmb.com/how-long-does-it-take-garbage-to-decompose-2878033. Accessed July 2018.

"Legal Limits on Single-Use Plastics and Microplastics: A Global Review of National Laws and Regulations." *United Nations Environment Programme*, https://wedocs.unep.org/bitstream/handle/20.500.11822/27113/plastics_limits.pdf?sequence=1&isAllowed=y. Accessed July 2023.

Parker, Laura. "A whopping 91% of plastic isn't recycled." *National Geographic News*, July 2017, https://news.nationalgeographic.com/2017/07/plastic-produced-recycling-waste-ocean-trash-debris-environment/. Accessed July 2018.

Professor Plastics. "What are Plastics?" *Plastics Make it Possible*, 10 June 2011, https://www.plasticsmakeitpossible.com/about-plastics/types-of-plastics/what-are-plastics/. Accessed July 2018.

"The New Plastics Economy: Rethinking the future of plastics." *World Economic Forum*, Jan 2016, http://www3.weforum.org/docs/WEF_The_New_Plastics_Economy.pdf. Accessed July 2018.

"'Turn the tide on plastic' urges UN, as microplastics in the seas now outnumber stars in our galaxy." *UN News*, 23 Feb 2017, https://news.un.org/en/story/2017/02/552052-turn-tide-plastic-urges-un-microplastics-seas-now-outnumber-stars-our-galaxy. Accessed July 2018.

"What's the problem with plastic bags." *The World Counts*, 15 April 2014, http://www.theworldcounts.com/stories/interesting-facts-about-plastic-bags. Accessed July 2018.

"22 Facts About Plastic Pollution (And 10 Things We Can Do About It)." *EcoWatch*, 7 April 2014, https://www.ecowatch.com/22-facts-about-plastic-pollution-and-10-things-we-can-do-about-it-1881885971.html. Accessed July 2018.

Breath of Fresh Air

"Air Quality: frequently asked questions." *Government of Canada,*
https://www.canada.ca/en/environment-climate-change/services/air-quality-health-ind
ex/frequently-asked-questions.html. Accessed August 2018.

"How to use the Air Quality Health Index." *Government of Canada,* 5 Feb 2016,
https://www.canada.ca/en/environment-climate-change/services/air-quality-health-ind
ex/use.html. Accessed August 2018.

The World Air Quality Project. "Air Quality Index Scale and Color Legend."
AQCIN.org, http://www.aqicn.org/scale/. Accessed August 2018.

Unraveling Your DNA

Ellen, Barbara. "What I learned from home DNA testing." *The Guardian*, 23 July
2017,
https://www.theguardian.com/science/2017/jul/23/what-i-learned-from-home-dna-te
st-kits-are-they-accurate-or-worthwhile. Accessed April 2018.

Hatfield, Heather. "Home DNA Tests: Buyer Beware." *WebMD*, 4 June 2007,
https://www.webmd.com/a-to-z-guides/features/home-dna-tests-buyer-beware.
Accessed April 2018.

"How it Works." *23andMe Canada,* https://www.23andme.com/en-ca/howitworks/.
Accessed April 2018.

NIH: U.S. National Library of Medicine. "What is DNA?" *Genetic Home Reference:
Your Guide to Understanding Genetic Conditions,*
https://ghr.nlm.nih.gov/primer/basics/dna. Accessed April 2018.

Park, Alice. "DNA Testing Kits Are on Everyone's Holiday List. 5 Things to Know If
You Get One." *TIME,* 19 Dec 2017,
http://time.com/5063464/23andme-dna-ancestry-test. Accessed April 2018.

Rossen, Jeff and Lindsey Bomnin. "How well do home DNA kits really work? See
identical triplets try 3 of them." *TODAY*, 1 Dec 2017,
https://www.today.com/health/are-home-dna-kits-accurate-identical-triplets-try-3-the
m-t119472. Accessed April 2018.

Can Sunscreen Really Give You Cancer?

Agin, Patricia P., Ruble, Karen, Hermansky, Steven J., and Timothy J. McCarthy. "Rates of allergic sensitization and irritation to oxybenzone-containing sunscreen products: a quantitative meta-analysis of 64 exaggerated use studies." *Photodermatology, Photoimmunology & Photomedicine* vol. 24, 4 (2008): 211 - 217.

"Benzophenone & Related Compounds." *Campaign for Safe Cosmetics,* http://www.safecosmetics.org/get-the-facts/chemicals-of-concern/benzophenone/. Accessed May 2018.

Brand, Rhonda M, Pike, James, Wilson, Roselyn M, et al. " Sunscreens containing physical UV blockers can increase transdermal absorption of pesticides." *Toxicology and Industrial Health* vol. 19, 1 (2003): 9 - 16.

"Can the chemicals in sunscreen cause cancer? *Canadian Cancer Society,* http://www.cancer.ca/en/prevention-and-screening/reduce-cancer-risk/make-healthy-choices/be-sun-safe/can-the-chemicals-in-sunscreen-cause-cancer/?region=on. Accessed May 2018.

Darbre, P. D., Aljarrah, A., Miller, W. R., Coldham, N. G., Sauer, M. J., and G. S. Pope. "Concentrations of parabens in human breast tumours." *Journal of Applied Toxicology* vol. 24, 1 (2004): 5 -13.

"Debunking Melanoma Myths: Do Sunscreens Cause Cancer?" *MDedge*, 22 Nov 2016, http://www.mdedge.com/cutis/article/118756/melanoma/debunking-melanoma-myths-do-sunscreens-cause-cancer. Accessed May 2018.

DiNardo, J. C., and C. A. Downs. "Why Evaluate the Sunscreen Active Oxybenzone (Benzophenone-3) for Carcinogenicity and Reproductive Toxicology or Consider it Unsafe for Human Use?" *Dermatology Research* vol. 1, 1 (2019): 1 - 3. [To replace a citation from the original blog post that is no longer accessible].

Fediuk, Daryl J., Wang, Tao, Chen, Yufei, Parkinson, Fiona E., Namaka, Michael P., Simons, Keith J., Burczynski, Frank J., and Xiaochen Gu. "Tissue disposition of the insect repellent DEET and the sunscreen oxybenzone following intravenous and topical administration in rats." *Biopharmaceutics & Drug Disposition* vol. 32, 7 (2011): 369 - 379.

Hart, Mike MD. "Is Your Sunscreen Causing Cancer?" *HuffPost Canada*, 7 Aug 2013, https://www.huffpost.com/archive/ca/entry/is-your-sunscreen-causing-cancer_b_3280 578?ec_carp=1709008773498610338. Accessed May 2018.

"Homosalate." *Campaign for Safe Cosmetics*, https://www.safecosmetics.org/chemicals/homosalate/. Accessed May 2018.

Jiménez-Díaz, I., Molina-Molina, J. M., Zafras-Gómez, A., Ballesteros, O., Navalón, A., Real, M., Sáenz, J. M., Fernández, M. F., and N. Olea. "Simultaneous determination of the UV-filters benzyl salicylate, phenyl salicylate, octyl salicylate, homosalate, 3-(4-methylbenzylidene) camphor and 3-benzylidene camphor in human placental tissue by LC–MS/MS. Assessment of their *in vitro* endocrine activity." *Journal of Chromatography B* vol. 936 (2013): 80 - 87.

Kim, Sujin, and Kyungho Choi. "Occurrences, toxicities, and ecological risks of benzophenone-3, a common component of organic sunscreen products: A mini-review." *Environment International* vol. 70 (2014): 143 - 157.

Krause, M., Klit, A., Blomberg Jensen, M., Søeborg, T., Frederiksen, H., Schlumpf, M., Lichtensteiger, W., Skakkebaek, N. E., and K. T. Drzewiecki. "Sunscreens: are they beneficial for health? An overview of endocrine disrupting properties of UV-filters." *International Journal of Andrology* vol. 35, 3 (2012): 424 - 436. [To replace a citation from the original blog post that is no longer accessible].

Layton, Scott M. "UV Filters as Common Organic Water Contaminants: A Toxicological Study of Selected UV Filters on *Daphnia Magna*, A Monitoring Study of Selected Oklahoma Lakes, and The Development of an Undergraduate Endocrine Disruption Authentic Research Lab." *Faculty of the Graduate College of the Oklahoma State University*, Dec 2015, https://shareok.org/bitstream/handle/11244/48913/Layton_okstate_0664D_14369. pdf?sequence=1&isAllowed=y. Accessed May 2018.

Ma, Risheng, Cotton, Bea, Lichtensteiger, Walter, and Margret Schlumpf. "UV Filters with Antagonistic Action at Androgen Receptors in the MDA-kb2 Cell Transcriptional-Activation Assay." *Toxicological Sciences* vol. 74, 1 (2003): 43 - 50.

MacLeman, Elle. "Avobenzone: What Do You Need To Know About Avobenzone Before You Use It?" *The Derm Review*, https://thedermreview.com/avobenzone/. Accessed May 2018.

MacLeman, Elle. "How Does Zinc Oxide Benefit Your Skin?" *The Derm Review,* https://thedermreview.com/zinc-oxide/. Accessed May 2018.

MacLeman, Elle. "Octocrylene: Why Octocrylene Use In Sunscreen?" *The Derm Review*, https://thedermreview.com/octocrylene/. Accessed May 2018.

Morison, Warwick L., and Steve Q. Wang. "Sunscreens: Safe and Effective?" *Skin Cancer Foundation*, https://www.skincancer.org/prevention/sun-protection/sunscreen/sunscreens-safe-and-effective. Accessed May 2018.

"Octisalate." *EWG's Skin Deep: Cosmetics Database*, http://www.ewg.org/skindeep/ingredient/704204/OCTISALATE/. Accessed May 2018.

"Opinion on Safety of Nanomaterials in Cosmetic Products." *Scientific Committee on Consumer Products*, http://ec.europa.eu/health/ph_risk/committees/04_sccp/docs/sccp_o_123.pdf. Accessed May 2018.

Pont, Adam R, Charron, Anna R, and Rhonda M Brand. "Active ingredients in sunscreens act as topical penetration enhancers for the herbicide 2,4-dichlorophenoxyacetic acid." *Toxicology and Applied Pharmacology* vol. 195, 3 (2004): 348 -354.

Schlumpf, Margret, Kypke, Karin, Wittassek, Matthias, Angerer, Juergen, Mascher, Hermann, Mascher, Daniel, Vökt, Cora, Birchler, Monika, and Walter Lichtensteiger. "Exposure patterns of UV filters, fragrances, parabens, phthalates, organochlor pesticides, PBDEs, and PCBs in human milk: Correlation of UV filters with use of cosmetics." *Chemosphere* vol. 81, 10 (2010): 1171 - 1183.

Schreurs, Richard, Lanser, Peter, Seinen, Willem, and Bart van der Burg. "Estrogenic activity of UV filters determined by an in vitro reporter gene assay and an in vivo transgenic zebrafish assay." *Toxicology* vol. 76, 5-6 (2002): 257 - 261.

Tavares, Renata S., Martins, Fátima C., Oliveira, Paulo J., Ramalho-Santos, João, and Francisco P. Peixoto. "Parabens in male infertility - Is there a mitochondrial connection?" *Reproduction Toxicology* vol. 27, 1 (2009): 1 - 7.

"Titanium Dioxide." *Campaign for Safe Cosmetics*, http://www.safecosmetics.org/get-the-facts/chemicals-of-concern/titanium-dioxide-2/. Accessed May 2018.

Walters, K. A., Brain, K. R., Howes, D., James, V. J., Kraus, A. L., Teetsel, N. M., Toulon, M., Watkinson, A. C., and S. D. Gettings. "Percutaneous penetration of octyl salicylate from representative sunscreen formulations through human skin *in vitro*" *Food and Chemical Toxicology* vol. 35, 12 (1997): 1219 -1225.

World Health Organization. "Benzophenone." *International Agency for Research on Cancer Monographs on the Identification of Carcinogenic Hazards to Humans* vol. 101, 7: https://monographs.iarc.fr/ENG/Monographs/vol101/mono101-007.pdf. Accessed May 2018.

Weisbrod, Christin J., Kunz, Petra Y., Zenker, Armin K., and Karl Fent. "Effects of the UV filter benzophenone-2 on reproduction in fish." *Toxicology and Applied Pharmacology* vol. 225, 3 (2007): 255 - 266.

What the SPF?

"Ask the Expert: Does a High SPF Protect My Skin Better?" *Skin Cancer Foundation*, 24 May 2018. https://www.skincancer.org/skin-cancer-information/ask-the-experts/does-a-higher-spf-sunscreen-always-protect-your-skin-better. Accessed June 2018.

Vegan vs. Meatarian: The Science of Food

Agency. "A third of 'vegetarians' eat meat when drunk on a night out." *The Telegraph,* 7 Oct 2015, https://www.telegraph.co.uk/news/uknews/11918091/A-third-of-vegetarians-eat-meat-when-drunk-on-a-night-out.html. Accessed July 2018.

Briffa, John, MD. "It's not just the salt that makes many meat-substitute foods a thoroughly unhealthy option." *Dr. Briffa*, 19 May 2018, http://www.drbriffa.com/2008/05/19/its-not-just-the-salt-that-makes-many-meat-substitute-foods-a-thoroughly-unhealthy-option/. Accessed July 2018.

Briffa, John, MD. "When is a 'healthy' food not a healthy food after all?" *Dr. Briffa,* 18 April 2018, http://www.drbriffa.com/2008/04/18/when-is-a-healthy-food-not-a-healthy-food-after-all. Accessed July 2018.

El Tinay, A. H., Mahgoub, S. O., Mohamed, B. E., and M. A. Hamad. "Proximate composition and mineral and phytate contents of legumes grown in Sudan." *Journal of Food Composition and Analysis* vol. 2, 1 (1989): 69 -78.

Gelsomin, Emily. "Impossible and Beyond: How healthy are these meatless burgers?" *Harvard Health Publishing*, 24 January 2022, https://www.health.harvard.edu/blog/impossible-and-beyond-how-healthy-are-these-meatless-burgers-2019081517448. Accessed July 2023.

Harvard T. H. Chan: School of Public Health. "Salt and Sodium." *The Nutrition Source,* https://www.hsph.harvard.edu/nutritionsource/salt-and-sodium/. Accessed July 2018.

Haupt, Angela. "Ashton Kutcher's Fruitarian Diet: What Went Wrong?" *U.S. News: Health*, 7 Feb 2013, https://health.usnews.com/health-news/articles/2013/02/07/ashton-kutchers-fruitarian-diet-what-went-wrong. Accessed July 2018.

Kindy, Kimberly. "Parents' lawsuit says Quorn mold-based food product killed their 11-year-old son." *The Washington Post*, 25 March 2015, https://www.washingtonpost.com/news/federal-eye/wp/2015/03/25/parents-lawsuit-says-quorn-mold-based-food-product-killed-their-11-year-old-son/?noredirect=on. Accessed July 2018.

Nöthlings, Ute, Murphy, Suzanne P., Wilkens, Lynne R., Henderson, Brian E., and Laurence N. Kolonel. "Dietary glycemic load, added sugars, and carbohydrates as risk factors for pancreatic cancer: the Multiethnic Cohort Study 1'2'3'4." *The American Journal of Clinical Nutrition* vol. 86, 5 (2007): 1495 - 1501.

Poulter, Sean. "Veggie sausages contain FIVE TIMES as much salt as a bag of crisps." *Daily Mail Online,* 19 May 2008, http://www.dailymail.co.uk/health/article-567144/Veggie-sausages-contain-FIVE-TIMES-salt-bag-crisps.html. Accessed July 2018.

"Questions and Answers - Hormonal Growth Promoters." *Government of Canada: Health Canada,* 5 Sept 2012, https://www.canada.ca/en/health-canada/services/drugs-health-products/veterinary-drugs/factsheets-faq/hormonal-growth-promoters.html. Accessed July 2018.

"Quorn." *Center for Science in the Public Interest,* https://cspinet.org/eating-healthy/foods-avoid/quorn. Accessed July 2018.

"The truth about carbs." *NHS,* https://www.nhs.uk/live-well/healthy-weight/why-we-need-to-eat-carbs/. Accessed July 2018.

Ying, Foo Jie, "Going Vegan Vs Eating Meat: Which Is Really Healthier?" *Men's Health,* 1 Nov 2017, http://www.menshealth.com.sg/weight-loss-nutrition/going-vegan-vs-eating-meat-which-really-healthier/. Accessed July 2018.

Big Pharma's Big Cover Up? A Look At The Rife Machine As A Cure For Cancer

"BACILLI REVEALED BY NEW MICROSCOPE; Dr. Rife's Apparatus, Magnifying 17,000 Times, Shows Germs Never Before Seen." *The New York Times,* 22 Nov 1931, https://www.nytimes.com/1931/11/22/archives/bacilli-revealed-by-new-microscope-dr-rifes-apparatus-magnifying.html. Accessed Sept 2018.

Barrett, Stephen, MD. "Rife Machine Operator Sued." *Quackwatch,* 10 Dec 2012, https://www.quackwatch.org/04ConsumerEducation/News/rife.html. Accessed Sept 2018.

"Lifetime Risk of Developing or Dying From Cancer." *American Cancer Society,* 4 Jan 2018, https://www.cancer.org/cancer/cancer-basics/lifetime-probability-of-developing-or-dying-from-cancer.html. Accessed Sept 2018.

Mac Manus, M. P., "Unproven medical devices and cancer therapy: big claims but no evidence." *Biomedical Imaging and Intervention Journal* vol. 4, 3 (2008): e25.

Rife, Royal Raymond. "History of the Development of a Successful. Treatment for Cancer and Other Virus, Bacteria and Fungi." *True Rife,* http://truerife.com/RifeBXDOC.pdf. Accessed Sept 2018.

The ALSUntangled Group. "ALSUntangled No. 23: The Rife Machine and retroviruses." *Amyotrophic Lateral Sclerosis and Frontotemporal Degeneration* vol. 15, 1 - 2 (2014): 157 -159.

Truman Jr., Stanley. *Rife.org,* https://rife.org/. Accessed Sept 2018.

The Not-So-Secret Side of Dream Science

Blackmore, Susan. "What happens when we dream?" *Science Focus,* https://www.sciencefocus.com/the-human-body/what-happens-when-we-dream/. Accessed Sept 2018.

"Dreams and their relation to physical and mental well-being." *The Free Library,* https://www.thefreelibrary.com/Dreams+and+their+relation+to+physical+and+mental+well-being.-a0272670908. Accessed Sept 2018.

"Facts About Dreaming." *WebMD,* https://www.webmd.com/sleep-disorders/guide/dreaming-overview. Accessed Sept 2018.

Nir, Yuval, and Giulio Tononi. "Dreaming and the brain: from phenomenology to neurophysiology." *Trends in Cognitive Sciences* vol. 14, 2 (2010): 88 - 100.

Rechtschaffen, Allan, Bergmann, Bernard M., Everson, Carol A., Kushida, Clete A., and Marcia A. Gilliland. "Sleep Deprivation in the Rat." *Sleep* vol. 12, 1 (1989): 1 - 67.

"Stages of Sleep and Sleep Cycles." *Tuck*, https://www.tuck.com/stages/. Accessed Sept 2018.

Cuppa Poison?

"Allowed and Prohibited Substances." *USDA: Agriculture Marketing Service,* https://www.ams.usda.gov/publications/content/allowed-prohibited-substances. Accessed July 2023.

"Blended With Care: From Seed to Sip." *Celestial Seasonings,* https://celestialseasonings.com/pages/blended-with-care-from-seed-to-sip. Accessed July 2023.

"BRCGS Food Safety." *Intertek,* https://www.intertek.com/assurance/brc/. Accessed July 2023.

"Brooke Bond." *RateTea.com*, 25 January 2018, https://ratetea.com/brand/brooke-bond/234/. Accessed July 2023.

Christiansen, A., Gervais, J., Buhl, K., and D. Stone. "Acephate General Fact Sheet." *National Pesticide Information Center, Oregon State University Extension Services,* 2011, http://npic.orst.edu/factsheets/acephagen.html. Accessed Nov 2018.

"Corporate Social Responsibility." *Goodricke: The Tea People,* https://www.goodricke.com/csr. Accessed July 2023.

Fraser, Carly. "8 Popular Tea Bag Companies That Contain Illegal Amounts of Deadly Pesticides." *Live Love Fruit,* 26 Oct 2016, https://livelovefruit.com/8-popular-tea-bag-companies-contain-illegal-amounts-deadly-pesticides/. Accessed Nov 2018.

"For Better Sourcing." *Tata Consumer Products,* https://www.tataconsumer.com/sustainability/better-sourcing. Accessed July 2023.

"Glaucus Research Group California LLC is initiating coverage on the Hain Celestial Group (Nasdaq: HAIN) with a Strong Sell rating." *Glaucus Research Group*, 21 Feb 2013, https://glaucusresearch.com/wp-content/uploads/downloads/2013/02/GlaucusResearch-The_Hain_Celestial_Group_Inc-NasdaqHAIN-Strong_Sell_Febuary_21_2013.pdf. Accessed Nov 2018.

Griffith-Greene, Megan. "Pesticide traces in some tea exceed allowable limits." *CBC News*, 8 March 2014, https://www.cbc.ca/news/canada/pesticide-traces-in-some-tea-exceed-allowable-limits-1.2564624. Accessed Nov 2018.

GreenPeace India. "Wagh Bakri commits to steps towards elimination of pesticides from tea cultivation. 40,000 citizens sign the petition asking tea companies to clean chai." *GreenPeace,* 25 September 2014, https://www.greenpeace.org/india/en/press/2484/wagh-bakri-commits-to-steps-towards-elimination-of-pesticides-from-tea-cultivation-40000-citizens-sign-the-petition-asking-tea-companies-to-clean-chai/. Accessed July 2023.

"Gujarat Tea Processors and Packers Ltd." *BRCGS Directory*, https://directory.brcgs.com/site/1987850. Accessed July 2023.

"How Can We Help You?" *Lipton*, https://www.lipton.com/us/en/faqs/. Accessed July 2023.

National Center for Biotechnology Information. "Acetamiprid." *PubChem,* 2011, https://pubchem.ncbi.nlm.nih.gov/compound/Acetamiprid. Accessed Nov 2018.

National Center for Biotechnology Information. "Actara." *PubChem*, https://pubchem.ncbi.nlm.nih.gov/compound/5821911. Accessed Nov 2018.

National Center for Biotechnology Information. "Bifenthrin." *PubChem*, https://pubchem.ncbi.nlm.nih.gov/compound/6442842. Accessed Nov 2018.

National Center for Biotechnology Information. "Buprofezin." *PubChem*, https://pubchem.ncbi.nlm.nih.gov/compound/50367. Accessed Nov 2018.

National Center for Biotechnology Information. "Carbendazim." *PubChem*, https://pubchem.ncbi.nlm.nih.gov/compound/25429. Accessed Nov 2018.

National Center for Biotechnology Information. "Chlorfenapyr." *PubChem*, https://pubchem.ncbi.nlm.nih.gov/compound/91778. Accessed Nov 2018.

National Center for Biotechnology Information. "Chlorpyrifos." *PubChem*, https://pubchem.ncbi.nlm.nih.gov/compound/2730. Accessed Nov 2018.

National Center for Biotechnology Information. "Clofentezine." *PubChem*, https://pubchem.ncbi.nlm.nih.gov/compound/73670. Accessed Nov 2018.

National Center for Biotechnology Information. "Cypermethrin." *PubChem*, https://pubchem.ncbi.nlm.nih.gov/compound/2912. Accessed Nov 2018.

National Center for Biotechnology Information. "Deltamethrin." *PubChem*, https://pubchem.ncbi.nlm.nih.gov/compound/40585. Accessed Nov 2018.

National Center for Biotechnology Information. "Dicofol." *PubChem*, https://pubchem.ncbi.nlm.nih.gov/compound/8268. Accessed Nov 2018.

National Center for Biotechnology Information. "Dimethoate." *PubChem*, https://pubchem.ncbi.nlm.nih.gov/compound/3082. Accessed Nov 2018.

National Center for Biotechnology Information. "Endosulfan." *PubChem*, https://pubchem.ncbi.nlm.nih.gov/compound/3224. Accessed Nov 2018.

National Center for Biotechnology Information. "Etofenprox." *PubChem*, https://pubchem.ncbi.nlm.nih.gov/compound/71245. Accessed Nov 2018.

National Center for Biotechnology Information. "Fenazaquin." *PubChem*, https://pubchem.ncbi.nlm.nih.gov/compound/86356. Accessed Nov 2018.

National Center for Biotechnology Information. "Fenpropathrin." *PubChem*, https://pubchem.ncbi.nlm.nih.gov/compound/47326. Accessed Nov 2018.

National Center for Biotechnology Information. "Fenpyroximate." *PubChem*, https://pubchem.ncbi.nlm.nih.gov/compound/9576412. Accessed Nov 2018.

National Center for Biotechnology Information. "Fipronil." *PubChem*, https://pubchem.ncbi.nlm.nih.gov/compound/3352. Accessed Nov 2018.

National Center for Biotechnology Information. "gamma-Cyhalothrin." *PubChem*, https://pubchem.ncbi.nlm.nih.gov/compound/6440554. Accessed Nov 2018.

National Center for Biotechnology Information. "Hexythiazox." *PubChem*, https://pubchem.ncbi.nlm.nih.gov/compound/13218777. Accessed Nov 2018.

National Center for Biotechnology Information. "Lannate." *PubChem*,
https://pubchem.ncbi.nlm.nih.gov/compound/5353758. Accessed Nov 2018.

National Center for Biotechnology Information. "Monocrotophos." *PubChem*,
https://pubchem.ncbi.nlm.nih.gov/compound/5371562. Accessed Nov 2018.

National Center for Biotechnology Information.
"N-[1-[(6-chloro-3-pyridinyl)methyl]-4,5-dihydroimidazol-2-yl]nitramide." *PubChem*,
https://pubchem.ncbi.nlm.nih.gov/compound/86418. Accessed Nov 2018.

National Center for Biotechnology Information. "Omethoate." *PubChem*,
https://pubchem.ncbi.nlm.nih.gov/compound/14210. Accessed Nov 2018.

National Center for Biotechnology Information. "Propargite." *PubChem*,
https://pubchem.ncbi.nlm.nih.gov/compound/4936. Accessed Nov 2018.

National Center for Biotechnology Information. "Pyridaben." *PubChem*,
https://pubchem.ncbi.nlm.nih.gov/compound/91754. Accessed Nov 2018.

National Center for Biotechnology Information. "Thiacloprid." *PubChem*,
https://pubchem.ncbi.nlm.nih.gov/compound/115224. Accessed Nov 2018.

National Center for Biotechnology Information. "Thiodicarb." *PubChem*,
https://pubchem.ncbi.nlm.nih.gov/compound/9601227. Accessed Nov 2018.

National Center for Biotechnology Information. "Triazophos." *PubChem*,
https://pubchem.ncbi.nlm.nih.gov/compound/32184. Accessed Nov 2018.

"Our Products." *Girnar: My Chai My Time,* https://girnar.com/our-products/.
Accessed July 2023.

"Quality." *Kanan Devan Hills Plantations Company (P) Limited,*
https://kdhptea.com/quality/. Accessed July 2023.

"Quality Certifications." *Wagh Bakri Tea Group,*
https://waghbakritea.com/m/certifications.php. Accessed July 2023.

"Search Products: Official QAI Product Certification Listing." *Quality Assurance
International (QAI),*
https://www.qai-inc.com/search-products/listing-detail.php?cust_id=107434&reques

tFrom=%2Flistings%2Flistings%5Fresults%5Fcompany%2Easp%3Flist%5Fby%3Du.
Accessed July 2023.

"TROUBLE BREWING: Pesticide residues in tea samples from India." *Greenpeace,*
2014,
https://www.greenpeace.org/archive-india/Global/india/image/2014/cocktail/downlo
ad/TroubleBrewing.pdf. Accessed Nov 2018.

University of Hertfordshire. "Beta-endosulfan." *PPDB: Pesticide Properties DataBase,*
11 June 2018, http://sitem.herts.ac.uk/aeru/ppdb/en/Reports/3028.htm. Accessed
Nov 2018.

Paper Plane Physics

"Paper Airplanes." *National Aeronautics and Space Administration*, 5 May 2015,
https://www.grc.nasa.gov/WWW/K-12/airplane/glidpaper.html. Accessed May 2018.

"What is Drag?" *National Aeronautics and Space Administration*, 5 May 2015,
https://www.grc.nasa.gov/www/k-12/airplane/drag1.html. Accessed May 2018.

"What is Lift?" *National Aeronautics and Space Administration*, 5 May 2015,
https://www.grc.nasa.gov/WWW/K-12/airplane/lift1.html. Accessed May 2018.

"What is Thrust?" *National Aeronautics and Space Administration*, 5 May 2015,
https://www.grc.nasa.gov/WWW/K-12/airplane/thrust1.html. Accessed May 2018.

"What is Weight?" *National Aeronautics and Space Administration*, 5 May 2015,
https://www.grc.nasa.gov/www/k-12/airplane/weight1.html. Accessed May 2018.

Luck of the Draw

Ossola, Alexandra. "The Science Of Luck." *Popular Science,* 17 March 2015,
https://www.popsci.com/luck-real/. Accessed June 2018.

Journey to Alpha Centauri: A Review of the Netflix Reboot – Lost in Space

Emspak, Jesse. "Is the Nearest Alien Planet Proxima b Habitable? 'It's Complicated.'" *Space.com, 31 Aug 2016,* https://www.space.com/33915-newfound-planet-proxima-b-habitability.html. Accessed April 2018.

Jackson, Mark. "Why Interstellar Travel Will Be Possible Sooner Than You Think." *Singularity Hub,* 18 June 2017, https://singularityhub.com/2017/06/18/why-interstellar-travel-will-be-possible-sooner-than-you-think/. Accessed April 2018.

National Aeronautics and Space Administration: Goddard Space Flight Center. "The Cosmic Distance Scale." *Imagine The Universe!,* 06 Dec 2016, https://imagine.gsfc.nasa.gov/features/cosmic/nearest_star_info.html. Accessed April 2018.

"Parker Solar Probe." *NASA Science Beta*, 22 May 2016, https://science.nasa.gov/missions/solar-probe. Accessed April 2018.

Scharf, Caleb A. "The Fastest Spacecraft Ever?"*Scientific American: Life, Unbounded,* 25 Feb 2013, https://blogs.scientificamerican.com/life-unbounded/the-fastest-spacecraft-ever/. Accessed April 2018.

George of the Urban Jungle: A Look At The Movie Rampage

"Acromegaly." *National Institute of Diabetes and Digestive and Kidney Diseases,* April 2012, https://www.niddk.nih.gov/health-information/endocrine-diseases/acromegaly. Accessed April 2018.

"Adam Rainer: The Giant and the Dwarf." *Historic Mysteries*, 13 May 2017, https://www.historicmysteries.com/adam-rainer-dwarf-giant/. Accessed April 2018.

"AquAdvantage Salmon Fact Sheet." *FDA,* https://www.fda.gov/animal-veterinary/aquadvantage-salmon/aquadvantage-salmon-fact-sheet. Accessed April 2018.

Ballantyne, Coco. "What causes albinism?" *Scientific American: Health*, 18 Feb 2009, https://www.scientificamerican.com/article/killing-albinos-tanzania-albinism/. Accessed April 2018.

Clancy, Suzanne. "RNA Functions." *Nature Education* vol. 1, 1 (2008): 102.

Didymus, Johnthomas. "Snowflake, famous albino gorilla was inbred, DNA sequencing shows." *Digital Journal*, 18 June 2013, http://www.digitaljournal.com/article/352581. Accessed April 2018.

Fero, Matthew L., Rivkin, Michael, Tasch, Michael, Porter, Peggy, Carow, Catherine E., Firpo, Eduardo, Polyak, Kornelia, Tsai, Li-Huei, Broudy, Virginia, Perlmutter, Roger M., Kaushansky, Kenneth, and James M Roberts. "A Syndrome of Multiorgan Hyperplasia with Feature of Gigantism, Tumorigenesis, and Female Sterility in p27^{Kip1}-Deficient Mice." *Cell* vol. 85, 5 (1996): 733 - 744.

"Gigantism." *MedlinePlus*, https://medlineplus.gov/ency/article/001174.htm. Accessed April 2018.

"GORILLA FACT SHEET." *World Animal Foundation*, http://www.worldanimalfoundation.net/f/Gorilla.pdf. Accessed April 2018.

"How cancer starts." *Cancer Research UK*, 23 Nov 2017, https://www.cancerresearchuk.org/about-cancer/what-is-cancer/how-cancer-starts. Accessed April 2018.

Miller, John. "All About Albinism." *Missouri Department of Conservation | Missouri Conservationist Magazine,* 17 Nov 2010, https://mdc.mo.gov/conmag/2005/06/all-about-albinism. Accessed April 2018.

Moss, Laura. "12 bizarre examples of genetic engineering." *MNN.com,* 27 Oct 2010, https://www.mnn.com/green-tech/research-innovations/photos/12-bizarre-examples-of-genetic-engineering/fast-growing. Accessed April 2018.

"Technology." *AquaBounty,* https://aquabounty.com/innovation/technology/. Accessed April 2018.

"Understanding the Immune System: How It Works." *U.S. Department of Health and Human Services, National Institutes of Health*, Sept 2003, http://www.imgt.org/IMGTeducation/Tutorials/ImmuneSystem/UK/the_immune_system.pdf. Accessed April 2018.

"What Is Marfan Syndrome?" *The MarFan Foundation, 2014,* https://www.marfan.org/about/marfan. Accessed April 2018.

Technology Behind Avengers: Infinity War

Carlyle, Ryan. "How Iron Man's Arc Reactor (Probably) Works." *Gizmodo*, 21 Nov 2014, https://gizmodo.com/how-iron-mans-arc-reactor-probably-works-1661613682. Accessed May 2018.

Dunning, J. J. "This amazing real-life Iron Man suit really flies." *Red Bull*, 4 Oct 2017, https://www.redbull.com/ie-en/real-life-iron-man-suit. Accessed May 2018.

"Jasper"- http://jasperproject.github.io/

Medina, Pat. "The US military plans to release a "supersoldier" suit in 2018." *Futurism*, 9 Oct 2015, https://futurism.com/the-us-military-plans-to-release-a-supersoldier-suit-in-2018. Accessed May 2018.

Tegler, Eric. "Exoskeletons won't turn assembly line workers into Iron Man." *ars technica*, 20 Nov 2017, https://arstechnica.com/cars/2017/11/exoskeletons-wont-turn-assembly-line-workers-into-iron-man/. Accessed May 2018.

Woollaston, Victoria. "Jarvis is real! Mark Zuckerberg has spent 2016 building an Iron Man-style AI that controls his life." *Wired*, 19 Dec 2016, http://www.wired.co.uk/article/mark-zuckerberg-jarvis-ai. Accessed May 2018.

Finding Mermaids: Analyzing Mermaids: A Body Found

"Cave of Swimmers, Egypt." *The British Museum*, https://africanrockart.britishmuseum.org/country/egypt/cave-of-swimmers/. Accessed May 2018.

Fernández, A., Edwards, J. F., Rodríguez, F., Espinosa De Los Monteros, A., Herráez, P., Castro, P., Jaber, J. R., Martín, V., and M. Arbelo. "Gas and Fat Embolic Syndrome" Involving a Mass Stranding of Beaked Whales (Family *Ziphiidae)* Exposed to Anthropogenic Sonar Signals. *Veterinary Pathology* vol. 42, 4 (2016): 446 - 457.

Malakoff, David. "A Roaring Debate Over Ocean Noise." *Science* vol. 291, 5504 (2001): 576 - 578.

Nieukirk, Sharon L., Stafford, Kathleen M., Mellinger, David K., Dziak, Robert P., and Christopher G. Fox. "Low-frequency whale and seismic airgun sounds recorded in the mid-Atlantic Ocean." *Journal Acoustical Society of America* vol. 114, 4 (2004): 1832 - 1843.

"NOAA FISHERIES SERVICE RELEASES FINAL REPORT ON 2004 STRANDING OF MELON-HEADED WHALES IN HAWAII." *NOAA Magazine,* 27 April 2006, https://www.sos.wa.gov/library/newspaperindex.aspx?oit=Whale&oim=&oiy=2004&oid=&oib=ArticleDate. Accessed May 2018.

"What is the bloop?" *National Ocean Service: National Oceanic and Atmospheric Administration*, https://oceanservice.noaa.gov/facts/bloop.html. Accessed May 2018.

Genetic Memory: Looking at Assassin's Creed

Alvarado, Sebastian. "The Science Fact Animating *Assassin's Creed's* Animus." *Kotaku*, 11 April 2012, https://kotaku.com/5901160/the-science-fact-animating-assassins-creeds-animus. Accessed May 2018.

Dias, Brian G., and Kerry J Ressler. "Parental olfactory experience influences behavior and neural structure in subsequent generations." *Nature Neuroscience* vol. 17 (2014): 89 - 96.

Gallagher, James. "'Memories' pass between generations." *BBC News*, 1 Dec 2013, https://www.bbc.com/news/health-25156510. Accessed May 2018.

Nishimoto, Shinji, Vu, An T, Naselaris, Thomas, Benjamini, Yuval, Yu, Bin, and Jack L Gallant. "Reconstructing visual experiences from brain activity evoked by natural movies." *Current Biology* vol. 21, 19 (2011): 1641 - 1646.

"Virus Infections and Hosts." *Lumen Learning*, https://courses.lumenlearning.com/boundless-biology/chapter/virus-infections-and-hosts/. Accessed May 2018.

Walia, Arjun. "Researchers Discover That Memories Can Be Passed Down Through Changes In Our DNA." *Collective Evolution*, 12 Nov 2015, https://www.collective-evolution.com/2015/11/12/researchers-discover-that-memories-can-be-passed-down-through-changes-in-our-dna/. Accessed May 2018.

Zimmer, Carl. "Our Inner Viruses: Forty Million Years In the Making." *National Geographics*, 1 Feb 2015, https://www.nationalgeographic.com/science/article/our-inner-viruses-forty-million-years-in-the-making. Accessed May 2018.

Regrowing Deadpool

"Adult Stem Cells." *University of Notre Dame: Center for Stem Cells and Regenerative Medicine*, http://stemcell.nd.edu/research/alternative-stem-cell-sources/adult-stem-cells/. Accessed May 2018.

Bryner, Jeanna. "How Salamanders Sprout New Limbs." *LiveScience*, 1 Nov 2007, https://www.livescience.com/1985-salamanders-sprout-limbs.html. Accessed May 2018.

Dimitriou, Rozalia, Jones, Elena, McGonagle, Dennis, and Peter V Giannoudis. "Bone regeneration: current concepts and future directions." *BMC Medicine* vol. 9, 66 (2011).

Dye, Briana R., Hill, David R., Ferguson, Michael A. H., Tsai, Yu-Hwai, Nagy, Melinda S., Dyal, Rachel, Wells, James M., Mayhew, Christopher N., Nattiv, Roy, Klein, Ophir D., White, Eric S., Deutsch, Gail H., and Jason R. Spence. "In vitro generation of human pluripotent stem cell derived lung organoids." *eLife* vol. 4 (2015).

Emspak, Jesse. "Could Humans Ever Regenerate a Limb?" *LiveScience*, 22 May 2017, https://www.livescience.com/59194-could-humans-ever-regenerate-limbs.html. Accessed May 2018.

Ericson, John. "Study Reveals Body's Secret To Skin Regeneration Rejuvenation: What Fibroblast Cells Can Teach Us About Skin Health." *Medical Daily*, 12 Dec 2013, https://www.medicaldaily.com/study-reveals-bodys-secret-skin-regeneration-rejuvenation-what-fibroblast-cells-can-teach-us-about. Accessed May 2018.

Geggel, Laura. "Tiny Human Stomachs Grown in Lab." *LiveScience*, 29 Oct 2014, https://www.livescience.com/48519-miniature-human-stomachs.html. Accessed May 2018.

Gholipour, Bahar. "Lab-Grown Vaginas Implanted Successfully in 4 Teenagers." *Scientific American*, 11 April 2014, https://www.scientificamerican.com/article/lab-grown-vaginas-implanted-successfully-in-4-teenagers/. Accessed May 2018.

Illingworth, Cynthia M. "Trapped fingers and amputated finger tips in children." *Journal of Pediatric Surgery* vol. 9, 6 (1974): 853 - 858.

Kessler, M., Hoffman, K., Brinkmann, V., Thieck, O., Jackisch, S., Toelle, B., Berger, H., Mollenkopf, H. J., Mangler, M., Sehouli, J., Fotopoulou, C., and T. F. Meyer. "The Notch and Wnt pathways regulate stemness and differentiation in human fallopian tube organoids." *Nature Communications* vol. 6, 8989 (2015).

Lewis, Tanya. "Missing Parts? Salamander Regeneration Secret Revealed." *LiveScience*, 20 Mar 2013,

https://www.livescience.com/34513-how-salamanders-regenerate-lost-limbs.html.
Accessed May 2018.

"Liver Regeneration." *Mayo Clinic: Center for Regenerative Medicine*,
https://www.mayo.edu/research/centers-programs/center-regenerative-medicine/focus
-areas/liver-regeneration. Accessed May 2018.

Ma, Zhen, Wang, Jason, Loskill, Peter, Huebsch, Nathaniel, Koo, Sangmo, Svedlund,
Felicia L., Marks, Natalie C., Hua, Ethan W., Grigoropoulos, Costas P., Conklin, Bruce
R., and Kevin E. Healy. "Self-organizing human cardiac microchambers mediated by
geometric confinement." *Nature Communications* vol. 6, 7413 (2015).

Minoru, Takasato, Er, Pei X., Chiu, Han S., Maier, Barbara, Baillie, Gregory J.,
Ferguson, Charles, Parton, Robert G., Wolvetang, Ernst J., Roost, Matthias S., Chuva
de Sousa Lopes, Susana M., and Melissa H. Little. "Kidney organoids from human iPS
cells contain multiple lineages and model human nephrogenesis." *Nature* vol. 526
(2015): 564 - 568.

Okamoto, Ryuichi. "Epithelial regeneration in inflammatory bowel diseases."
Inflammation and Regeneration vol. 31, 3 (2011): 275 - 281.

"Scientist: Most complete human brain model to date is a 'brain changer.'" *The Ohio
State University News*, 18 Aug 2015,
https://news.osu.edu/scientist-most-complete-human-brain-model-to-date-is-a-brain-c
hanger/. Accessed May 2018.

"Stem Cell Basics I." *National Institute of Health*,
https://stemcells.nih.gov/info/basics/1.htm. Accessed May 2018.

"Valvular and Vascular Repair and Regeneration." *Mayo Clinic: Center for
Regenerative Medicine*,
https://www.mayo.edu/research/centers-programs/center-regenerative-medicine/focus
-areas/valvular-vascular-repair-regeneration. Accessed May 2018.

Racing the Millennium Falcon

Lough, Chris. "How Fast is the *Millennium Falcon*? A Thought Experiment."
TOR.COM, 8 Dec 2014,

https://www.tor.com/2014/12/08/star-wars-how-fast-is-the-millennium-falcon/.
Accessed May 2018.

"Millennium Falcon." *Wookieepedia*,
https://starwars.fandom.com/wiki/Millennium_Falcon. Accessed May 2018.

"The galaxy." *Wookieepedia*, https://starwars.fandom.com/wiki/The_galaxy. Accessed
May 2018.

Wallace, Daniel, and Jason Fry. *Star Wars: The Essential Atlas,* Random House
Worlds, 2009.

The Jurassic Park Question: 5 Extinct Animals Being Brought Back

"Bringing Australian Animals Back To Life." *National Geographic | Wild Australia*,
https://www.nationalgeographic.com.au/animals/bringing-australian-animals-back-to
-life.aspx. Accessed July 2018.

Choi, Charles Q. "First Extinct-Animal Clone Created." *National Geographic*, 10 Feb
2009,
https://www.nationalgeographic.com/science/2009/02/news-bucardo-pyrenean-ibex-
deextinction-cloning/. Accessed July 2018.

Kenneally, Christine. "A 'De-extinction' Company Wants to Bring Back the Dodo."
Scientific American, 31 January 2023,
https://www.scientificamerican.com/article/tech-company-invests-150m-to-bring-bac
k-the-dodo/. Accessed July 2023.

Lewis, Danny. "The Last Wooly Mammoths Died Isolate and Alone."
Smithsonian.com: Smart News, 8 May 2015,
https://www.smithsonianmag.com/smart-news/last-wooly-mammoths-died-isolated-a
nd-alone-180955208. Accessed July 2018.

Milman, Oliver. "Zoologists hunting Tasmanian tiger declare 'no doubt' species still
alive." *The Guardian,* 11 Nov 2013,
https://www.theguardian.com/world/2013/nov/11/zoologists-on-the-hunt-for-tasma
nian-tiger-declare-no-doubt-species-still-alive. Accessed July 2018.

Mueller, Tom. "Ice Baby." *National Geographic,* May 2009,
https://www.nationalgeographic.com/magazine/2009/05/mammoths/. Accessed July
2018.

Newcomb, Tim. "Scientists Are Hell-Bent on Resurrecting the Tasmanian Tiger. Here's Their Complicated Plan." *Popular Mechanics*, 22 August 2022, https://www.popularmechanics.com/science/animals/a40919781/de-extinction-bringing-back-tasmanian-tiger/. Accessed July 2023.

Newcomb, Tim. "Scientists Are Reincarnating the Woolly Mammoth to Return in 4 Years." *Popular Mechanics,* 30 January 2023, https://www.popularmechanics.com/science/animals/a42708517/scientists-reincarnating-woolly-mammoth/. Accessed July 2023.

Searle, Adam. "Spectral ecologies: De/extinction in the Pyrenees." *Transactions of the Institute of British Geographers* vol 47, 1 (2021): 167-183.

Strauss, Bob. "10 Facts About the Tasmanian Tiger." *Thought Co.,* 6 Feb 2019, https://www.thoughtco.com/facts-about-the-tasmanian-tiger-1093338. Accessed July 2018.

TED. "Stewart Brand: The dawn of de-extinction. Are you ready?" *YouTube.com,* 13 March 2013, https://www.youtube.com/watch?v=XKc9MJDeqj0. Accessed July 2018.

The Quagga Project, https://www.quaggaproject.org/. Accessed July 2018 & July 2023.

"The Quagga Project." *Facebook,* https://www.facebook.com/quaggaproject. Accessed July 2018 & July 2023.

Dissecting Frankenstein's Monster

BEC Crew. "World's First Head Transplant Volunteer Could Experience Something 'Worse Than Death,'" *Science Alert*, 17 March 2018, https://www.sciencealert.com/world-s-first-head-transplant-volunteer-could-experience-something-worse-than-death. Accessed Oct 2018.

Canavero, Sergio. "Whole brain transplantation in man: Technically feasible." *Surgical Neurology International*, 23 December 2022, https://surgicalneurologyint.com/surgicalint-articles/whole-brain-transplantation-in-man-technically-feasible/. Accessed July 2023.

Colledge, Helen, and Elizabeth Boskey. "ABO Incompatibility Reaction." *Healthline*, 2 May 2017, https://www.healthline.com/health/abo-incompatibility. Accessed Oct 2018.

"Immunosuppressants." *National Kidney Foundation*, https://www.kidney.org/atoz/content/immuno. Accessed Oct 2018.

Kirkey, Sharon. "World's first human head transplant successfully performed on a corpse, scientists say." *National Post*, 17 November 2017, https://nationalpost.com/health/worlds-first-human-head-transplant-successfully-performed-on-a-corpse-scientists-say. Accessed July 2023.

Knoepfler, Paul. "Head transplant surgery update & fact check." *The Niche,* 7 January 2021, https://ipscell.com/2021/01/head-transplant-surgery-update-fact-check/. Accessed July 2023.

Mohney, Gillian. "'Frozen' Man Revived From the Brink of Death After Being Found in the Snow With No Pulse." *ABC News,* 19 Jan 2016, https://abcnews.go.com/Health/frozen-man-revived-brink-death-found-snow-pulse/story?id=36380318. Accessed Oct 2018.

Scott. "How Long A Person's Heart Has To Be Stopped Before Medics Wouldn't Try To Revive Them." *Today I Found Out*, 4 Nov 2013, http://www.todayifoundout.com/index.php/2013/11/long-persons-heart-stopped-wouldnt-try-revive/. Accessed Oct 2018.

Tapalaga, Andrei. "What Happened to the First Human Head Transplant?" *History of Yesterday*, 3 September 2022, https://historyofyesterday.com/what-happened-to-the-first-human-head-transplant/. Accessed July 2023.

"Tests you need before surgery." *Allina Health*, 15 May 2009, https://www.allinahealth.org/health-conditions-and-treatments/health-library/patient-education/kidney-transplant/for-the-kidney-donor/tests-you-need-before-surgery/. Accessed Oct 2018.

"The history of blood types." *Big Picture Education,* https://bigpictureeducation.com/history-blood-types. Accessed Oct 2018.

Nibiru: Destroyer of Earth

Dangerfield, Katie. "'Planet X' will destroy the Earth on April 23, according to doomsday prophecy." *Global News,* 15 April 2018, https://globalnews.ca/news/4142327/planet-x-nibiru-world-ending-april-23/. Accessed April 2018.

"Excerpted and edited from 'Interview with Zecharia Sitchin' from CONNECTING LINK issue 17." *Bibliotecapleyades.net,* https://www.bibliotecapleyades.net/sitchin/esp_sitchin_3a.htm. Accessed April 2018.

Fienberg, Richard Tresch. "What is a Sundog, and How Did "Sundogs" Get Their Name?" *Sky and Telescope*, 25 July 2006, https://www.skyandtelescope.com/astronomy-resources/astronomy-questions-answers/why-are-sundogs-called-by-that-name/. Accessed April 2018.

"Hypothetical Planet X." *NASA Science: Solar System Exploration*, https://solarsystem.nasa.gov/planets/hypothetical-planet-x/in-depth/. Accessed April 2018.

"Nancy Lieder's Bio." *ZetaTalk*, http://www.zetatalk.com/nancybio.htm. Accessed April 2018.

"Nibiru cataclysm." *Wikipedia*, https://en.wikipedia.org/wiki/Nibiru_cataclysm. Accessed April 2018.

Russell, Steve. "Will Nibiru Return in 2003? Interview With Zecharia Sitchin." *Bibliotecapleyades.net*, 1 June 2002, https://www.bibliotecapleyades.net/sitchin/esp_sitchin_0a.htm. Accessed April 2018.

Solar System Exploration Research Virtual Institute. "The Truth About Nibiru." *YouTube*, 21 Oct 2011, https://www.youtube.com/watch?v=1TIy-t48uU0. Accessed April 2018.

Sutherland, Scott. "Despite what you've heard, the world is NOT ending today." *The Weather Network*, 23 April 2018, https://www.theweathernetwork.com/news/articles/nibiru-hoax-rears-ugly-head-again-what-you-need-to-know-about-this-persistent-myth/99534. Accessed April 2018.

Wolchover, Natalie. "Believers In Mysterious Planet Nibiru Await Earth's End." *LiveScience*, 5 July 2011, https://www.livescience.com/33380-comet-elenin-planet-nibiru-doomsday-2012.html. Accessed April 2018. [Cross-posted on NASA's SERVI Solar System Exploration Research Virtual Institute: https://sservi.nasa.gov/articles/scientists-reject-impending-nibiru-earth-collision/]

Worrall, Simon. "Earth's Poles Will Eventually Flip, So What Then?" *National Geographic,* 1 February 2018, https://www.nationalgeographic.com/science/article/earth-magnetic-field-flip-poles-spinning-magnet-alanna-mitchell. Accessed July 2023.

"ZetaTalk: Communications." *ZetaTalk*, 15 Sept 1995, http://www.zetatalk.com/transfor/t18.htm. Accessed April 2018.

A Co-operative Mission of Firsts: InSight & Mars Cube One

"InSight Launch Press Kit." *NASA | Jet Propulsion Laboratory: California Institute of Technology*, https://www.jpl.nasa.gov/news/press_kits/insight/launch/download/mars_insight_launch_presskit.pdf. Accessed April 2018.

"InSight Mars Lander." *NASA*, https://www.nasa.gov/mission_pages/insight/overview/index.html. Accessed July 2023.

"InSight Top Science Results." *MARS InSight Mission*, https://mars.nasa.gov/insight/mission/science/results/. Accessed July 2023.

"MarCO (Mars Cube One)." *NASA Solar System Exploration*, https://solarsystem.nasa.gov/missions/mars-cube-one/in-depth/. Accessed July 2023.

"NASA InSight Study Provides Clearest Look Ever at Martian Core." *NASA: InSight Mars Lander,* 24 April 2023, https://www.nasa.gov/feature/jpl/nasa-insight-study-provides-clearest-look-ever-at-martian-core. Accessed July 2023.

"NASA Extends Exploration for 8 Planetary Science Missions." *NASA Science: MARS Exploration*, 25 April 2022, https://mars.nasa.gov/news/9175/nasa-extends-exploration-for-8-planetary-science-missions/. Accessed July 2023.

"NASA Retires InSight Mars Lander Mission After Years of Science." *MARS InSight Mission,* 21 December, 2022, https://mars.nasa.gov/news/9321/nasa-retires-insight-mars-lander-mission-after-years-of-science/?site=insight. Accessed July 2023.

Water on Mars?

Baxamusa, Batul Nafisa. "Archaebacteria Examples." *BiologyWise*, 8 April 2018, https://biologywise.com/archaebacteria-examples. Accessed July 2018.

"Cassini Finds Global Ocean in Saturn's Moon Enceladus." *NASA | Jet Propulsion Laboratory: California Institute of Technology News*, 15 Sept 2015, https://www.jpl.nasa.gov/news/news.php?feature=4718. Accessed July 2018.

Choi, Charles. "Water on Mars carved deep gullies and left a 'great puzzle' for Red Planet history." *Space.com*, 17 July 2023, https://www.space.com/mars-water-gullies-climate-change. Accessed July 2023.

Drake, Nadia. "Underground Lake Found on Mars? Get the Facts." *National Geographic*, 25 July 2018, https://www.nationalgeographic.com/science/2018/07/news-lake-found-mars-water-polar-cap-life-space/. Accessed July 2018.

Mitrofanov, I., Malakhov, A., Djachkova, M., Golovin, D., Litvak, M., Mokrousov, M., Sanin, A., Svedhem, H., and L. Zelenyi. "The evidence for unusually high hydrogen abundance in the central part of Valles Marineris on Mars." *Icarus,* vol. 374 (2022): https://www.sciencedirect.com/science/article/pii/S0019103521004528?via%3Dihub.

Strickland, Ashley. "'Significant amounts of water' found in Mars' massive version of the Grand Canyon." *CNN World: Space + Science,* 16 December 2021, https://www.cnn.com/2021/12/16/world/exomars-water-mars-grand-canyon-scn/index.html. Accessed July 2023.

Thompson, Jay R. "Why Europa: Evidence for an Ocean." *NASA: Europa Clipper*, https://europa.nasa.gov/why-europa/evidence-for-an-ocean/. Accessed July 2018.

Thompson, Joanna. "China's malfunctioning Mars rover may have found evidence of recent water on the Red Planet. *LiveScience,* 2 May 2023, https://www.livescience.com/space/mars/chinas-malfunctioning-mars-rover-may-have-found-evidence-of-recent-water-on-the-red-planet. Accessed July 2023.

Shooting For The Star: Looking at the Parker Solar Probe

Bale, S. D., Drake, J. F., McManus, M. D., Desai, M. I., Badman, S. T., Larson, D. E., Swisdak, M., Horbury, T. S., Raouafi, N. E., Phan, T., Velli, M., McComas, C. M. S., Mitchell, D., Panasenco, O., and J. C. Kasper. "Interchange reconnection as the source of the fast solar wind with coronal holes." *Nature*, vol. 618 (2023): 252-256.

"Eugene Newman Parker." *NASA: Parker Solar Probe,* https://www.nasa.gov/content/goddard/eugene-newman-parker. Accessed Aug 2018.

Cukier, W. Z., and J. R. Szalay. "Formation, Structure, and Detectability of the Geminids Meteoroid Stream." *The Planetary Science Journal*, vol. 4, 6 (2023): 109.

"Parker Scientists May Have Solved Decades-old Mysteries About the Origins of the Solar Wind." *NASA Parker Solar Probe,* 10 January 2023, http://parkersolarprobe.jhuapl.edu/News-Center/Show-Article.php?articleID=183. Accessed July 2023.

"Parker Solar Probe: The Mission." *The John Hopkins University Applied Physics Laboratory,* http://parkersolarprobe.jhuapl.edu/The-Mission/index.php. Accessed Aug 2018.

O'Regan, Alaina. "Scientists Shed Light on the Unusual Origin of a Familiar Meteoroid Stream." *NASA Parker Solar Probe,* 15 June 2023, http://parkersolarprobe.jhuapl.edu/News-Center/Show-Article.php?articleID=186. Accessed July 2023.

Sanders, Robert. "Parker Solar Probe Flies into Fast Solar Wind and Finds Its Source." *NASA Parker Solar Probe,* 8 June 2023, http://parkersolarprobe.jhuapl.edu/News-Center/Show-Article.php?articleID=185. Accessed July 2023.

"Solar storms." *Government of Canada | Canadian Space Agency*, 7 Aug 2017, https://www.asc-csa.gc.ca/eng/sciences/solar-storms.asp. Accessed Aug 2018.

"Spacecraft Status." *NASA Parker Solar Probe,* http://parkersolarprobe.jhuapl.edu/index.php. Accessed July 2023.

The Ultimate Guide to the Perseid Meteor Shower 2018

Lada, Brian. "Perseids to peak this weekend as one of 2018's best meteor showers." *AccuWeather*, 12 Aug 2018, https://www.accuweather.com/en/weather-news/clouds-wildfire-smoke-may-spoil-perseid-meteor-shower-for-many-this-weekend/70005720. Accessed Aug 2018.

Lewin, Sarah. "Comet Swift-Tuttle: The Icy Parent of the Perseid Meteor Shower." *Space.com*, 9 Aug 2016, https://www.space.com/33677-comet-swift-tuttle-perseid-meteor-shower-source.html. Accessed Aug 2018.

Making The World More Accessible, One Piece of Tech At A Time

Disabled World. "Disability Statistics: Information, Charts, Graphs and Tables." *Disabled World,* https://www.disabled-world.com/disability/statistics/. Accessed May 2018.

"GyroGlove." *GyroGear*, http://gyrogear.co/gyroglove. Accessed May 2018.

McGlashen, Keir. "How Gyroscopes Work." *YouTube*, 8 Oct 2015, https://www.youtube.com/watch?v=agVqyxSWxXs. Accessed May 2018.

Seth, Radhika. "Reading Braille Aloud." *Yanko Design*, 5 May 2011, https://www.yankodesign.com/2011/05/05/reading-braille-aloud/. Accessed May 2018.

"Understanding Cochlear Implants." *WebMD*, 1 June 2016, https://www.webmd.com/healthy-aging/understanding-cochlear-implants. Accessed May 2018.

Watch Where You're Driving!: A Look At Driverless Cars

ABC News. "Deadly crash with Tesla vehicle on auto pilot." *YouTube*, 31 March 2018, https://www.youtube.com/watch?v=VgQwHDFohTo. Accessed June 2018.

ABC News. "Woman dies after being hit by self-driving Uber." *YouTube*, 20 March 2018, https://www.youtube.com/watch?v=_3OBo0Nu3wU. Accessed June 2018.

AutomotiveTv. "Volvo Self Driving Car." *YouTube*, 2 Dec 2013, https://www.youtube.com/watch?v=bJwKuWz_lkE. Accessed June 2018.

"Autopilot." *Tesla Canada*, https://www.tesla.com/en_CA/autopilot?redirect=no. Accessed June 2018.

Boudette, Neal E. "Toyota Takes Self-Driving Cars Off Road After Uber Accident." *The New York Times*, 20 March 2018, https://www.nytimes.com/2018/03/20/business/uber-driverless-car-accident.html. Accessed June 2018.

Johnson, Leif, and Michelle Fitzsimmons. "Uber self-driving cars: everything you need to know." *TechRadar*, 25 May 2018, https://www.techradar.com/news/uber-self-driving-cars. Accessed June 2018.

KPIX CBS SF Bay Area. "Video Shows Tesla On Autopilot Nearly Crashing On Hwy 101." *YouTube*, 4 April 2018, https://www.youtube.com/watch?v=o02H2xGIecc. Accessed June 2018.

"Nissan's Self-Driving Car." *Nissan: Autonomous Car*, 25 May 2017, https://www.nissanusa.com/experience-nissan/news-and-events/self-driving-autonomous-car.html. Accessed June 2018.

Pasley, James. "The World's First Solar Road Has Officially Crumbled Into a Total Failure." *ScienceAlert*, 16 August 2019, https://www.sciencealert.com/the-world-s-first-solar-road-has-turned-out-to-be-a-disappointing-failure. Accessed July 2023.

Sanchez, Karla. "Daimler Successfully Demonstrates Self-Driving Big Rig." *MotorTrend*, 8 July 2014, https://www.motortrend.com/news/daimler-successfully-demonstrates-self-driving-big-rig/. Accessed June 2018.

"Self-Driving." *Lyft*, https://self-driving.lyft.com/. Accessed June 2018.

TheHUB. "How Do Self-Driving Cars Actually Work? (Tesla, Volvo, Google),"
YouTube, 17 Nov 2017, https://www.youtube.com/watch?v=xMH8dk9b3yA.
Accessed June 2018.

TomoNews US. "Tesla autopilot crash: Tesla Model S rear ends fire truck in Culver
City, California - TomoNews." *YouTube*, 24 Jan 2018,
https://www.youtube.com/watch?v=mEGAH330SCY&app=desktop. Accessed June
2018.

Valdes-Dapena, Peter. "Toyota reveals self-driving car." *CNN Business*, 7 Jan 2013,
https://money.cnn.com/2013/01/04/autos/toyota-self-driving-cars/index.html.
Accessed June 2018.

Waymo, https://waymo.com/. Accessed June 2018.

Zillman, Claire. "You Can Buy This Self-Driving Car for $20,000." *Fortune*, 14 March
2016, https://fortune.com/2016/03/14/self-driving-car-honda/. Accessed June 2018.

5 Canadian Innovations that Changed the World

Doetsch, Karl H., and Garry Lindberg. "Canadarm." *The Canadian Encyclopedia*, 4
March 2015, https://www.thecanadianencyclopedia.ca/en/article/canadarm. Accessed
July 2018.

Grant, Kelly. "A marvel of medicine near to Canada's heart." *The Globe and Mail*, 12
Nov 2017,
https://www.theglobeandmail.com/news/national/canada-150/the-pacemaker-is-a-mar
vel-of-medicine-near-to-canadasheart/article34859735/. Accessed July 2018.

Leung, Iris. "BlackBerry Limited." *The Canadian Encyclopedia*, 26 Feb 2012,
https://www.thecanadianencyclopedia.ca/en/article/blackberry-limited. Accessed July
2018.

Martin, Simon. "King man helped revolutionize movie restoration industry."
YorkRegion.com, 13 Nov 2013,
http://www.yorkregion.com/news-story/4207606-king-man-helped-revolutionize-mov
ie-restoration-industry/. Accessed July 2018.

"Our History." *Diabetes Canada,*
https://www.diabetes.ca/en-CA/about-diabetes-canada/our-history. Accessed July
2018.

"What is diabetes?" *Diabetes Canada,*
https://www.diabetes.ca/en-CA/diabetes-basics/what-is-diabetes. Accessed July 2018.

Driving on Sunshine: An Analysis of Solar Roads

"Average Sunshine a Year in British Columbia." *Current Results,*
https://www.currentresults.com/Weather/Canada/British-Columbia/sunshine-annual
-average.php. Accessed Sept 2018.

Economist Intelligence Unit. "US and Canada Green City Index: Assessing the
environmental performance of 27 major US and Canadian cities." *Siemens AG,* 2011,
https://www.siemens.com/entry/cc/features/greencityindex_international/all/en/pdf/r
eport_northamerica_en.pdf. Accessed Sept 2018

Finch, Graham, Burnett, Eric, and Warren Knowles. "Energy Consumption in Mid
and High Rise Residential Buildings in British Columbia." *BEST 2 - Energy Efficiency
- Session EE3-1,* https://www.brikbase.org/sites/default/files/EE3-1_finch_final_0.pdf.
Accessed Sept 2018. [To replace a citation from the original blog post that is no longer
accessible].

Intrinsik Corp. "Greenhouse Gas Emissions Associated with Various Methods of
Power Generation in Ontario." *Ontario Power Generation Inc.,* Oct 2016,
https://www.opg.com/darlington-refurbishment/Documents/IntrinsikReport_GHG
_OntarioPower.pdf. Accessed Sept 2018.

Johnstone, Patrick. "How Much Road Is Enough?" *Ask Pat,* 15 Aug 2014,
https://patrickjohnstone.ca/2014/08/how-much-road-is-enough.html. Accessed Sept
2018.

Osborn, Liz. "Yearly Snowfall Averages for Canadian Cities." *Current Results,*
https://www.currentresults.com/Weather/Canada/Cities/snowfall-annual-average.php
. Accessed Sept 2018.

"The Numbers." *SolarRoadways,* http://solarroadways.com/numbers.shtml. Accessed
Sept 2018.